CAD Books from OnWord Press

MicroStation 4.X
The MicroStation 4.X Delta Book
INSIDE MicroStation
INSIDE MicroStation Companion Workbook
INSIDE MicroStation Companion Workbook
    Instructor's Guide
MicroStation Reference Guide
MicroStation Productivity Book
101 MDL Commands
Bill Steinbock's Pocket MDL
    Programmer's Guide
MDL Guides
MicroStation For AutoCAD Users

MicroStation 3.X
INSIDE MicroStation
MicroStation Reference Guide
MicroStation Productivity Book

Other CAD Titles
The CAD Rating Guide
The One Minute CAD Manager

Books By Pen & Brush Publishers
Distributed by OnWord Press
The Complete Guide to MicroStation 3D
Programming With MDL
Programming With User Commands

MicroStation™ 4.X Delta Book

A Quick Guide For Upgrading to 4.X

By Frank Conforti

MicroStation™ 4.X Delta Book

A Quick Guide For Upgrading to 4.X

By Frank Conforti

Published by:

OnWord Press
P.O. Box 500
Chamisal, NM 87521 USA

All rights reserved. No part of this book may be reproduced or transmitted in any form or by any means, electronic or mechanical, including photocopying, recording or by any information storage and retrieval system without written permission from the publisher, except for the inclusion of brief quotations in a review.

Copyright © 1991 Frank Conforti

10 9 8 7 6 5 4 3 2

Printed in the United States of America

Library of Congress Cataloging-in-Publication Data

```
Frank Conforti
MicroStation™ 4.X Delta Book
A Quick Guide For Upgrading to 4.X

Includes index.

1. MicroStation (computer program)  I. Title
91-60274
ISBN 0-934605-34-3
```

Trademarks

MicroStation is a trademark of Bentley Systems, Inc. Intergraph is a registered trademark and IGDS is a trademark of Intergraph Corporation. OnWord Press is a trademark of DSR Publishing Company, Inc. Numerous additional products and services are mentioned in this book that are the trademarks or registered trademarks of their respective owners. Neither OnWord Press nor the author makes any claims to these marks.

Warning and Disclaimer

This book is designed to provide information about MicroStation. Every effort has been made to make this book complete and as accurate as possible; however, no warranty or fitness is implied.

The information is provided on an "as-is" basis. The author and OnWord Press shall have neither liability nor responsibility to any person or entity with respect to any loss or damages in connection with or rising from the information contained in this book.

About the Author

Frank Conforti has been involved in CAD since the mid 70's. He has worked with several CAD systems over the years. His involvement with Intergraph started in 1983 as a system manager for a VAX-based IGDS system in a CAD service bureau. During his tenure as system manager and later branch manager he had the opportunity to work with and train individuals from all different design disciplines.

Frank's involvement with Bentley Systems, Inc. began with Bentley's PseudoStation product, which successfully mated a Macintosh computer to the VAX IGDS system as a graphics workstation.

In 1988 Frank left the service bureau business to take on the position as MIS director for Keith and Schnars, P.A., a large south Florida civil engineering firm. At Keith and Schnars Frank has coordinated the ongoing effort to shift from VAX-based IGDS to PC- and Interpro-based MicroStation platforms.

Currently, when Frank is not busily writing books about CAD, he is traveling around the country conducting training seminars.

The author at home.

Frank estimates that, between all the CAD systems he has worked on, the new users he has initiated via his training number in the hundreds.

Frank is an avid and respected writer in the CAD field. He is a regular contributor of articles and reviews to Macintosh Aided Design, MacUser and MacWeek magazines. He also co-authored the Macintosh CAD/CAM book.

In his "spare" time, Frank tinkers with home automation hardware and software.

Thanks for the Help

I would like to thank Bentley Systems, Inc. for help and support and for providing a copy of MicroStation for each platform. Additional thanks goes to Scott Bentley, Keith Bentley, Barry Bentley and Keith Little for their not getting irritated at my odd requests over the years. See guys, it was all for the best!

I would also like to thank Dan Raker, David Talbott and Michelle Noel for their support and super effort in making this whole project work. Without their insistence on quality and their determination to make this book the best it can be, it would not have been possible.

I especially want to thank Dan for giving me the opportunity to pursue this dream (or nightmare, depending on which deadline he had me on!).

And to my special friends, who helped keep me "human" in between the "all nighters" and gallons of coffee required to write this (and other) books. Thanks guys!

And finally, I want to thank my number one fan who has been here through the whole project, my wife. Beccie, I know I couldn't have done this project without your unending support, your willingness to sacrifice your time to just lay low and keep me full of coffee.

For this reason I dedicate this book to you. Thank you dear.
    Frank Conforti
    February, 1991

Cover Art

Image by Jerry D. Flynn and Robert P. Humeniuk, Design Visualization Group, McDonnell Douglas Space Systems Company, Kennedy Space Center, Florida. The images were created using MicroStation and rendered using Intergraph Modelview and MicroStation 4.0.

A Sony GDM-1950 Monitor and Number Nine Computer 9GX level 3 video board was used for display of images. Originals were shot from screen using an Olympus OM4 35mm camera at F8 with Kodak Kodacolor 200 slide film.

The cover artist would especially like to thank Tony Clarey at Number Nine for the 9GX video board also Raymond Bentley and Brett Yeagley at Bentley Systems for their support during beta testing.

OnWord Press.....

OnWord Press is dedicated to the fine art of practical software user's documentation.

In addition to the authors who developed the material for this book, other members of the OnWord Press team make the book end up in your hands.

Thanks to: David Talbott, Michelle Noel, John Messeder, Dan Raker, Jean Nichols, and Sheila Miller.

Table of Contents

Introduction

Introduction
How This Book Is Organized xiii
    Command Illustrations xiv
    What Is Not Covered in this Book xv

Chapter 1

Welcome to MicroStation 4.0
A TOOL FOR THE NINETIES

The "New" Look 3
MicroStation Development Language 5
Back to Basics: the Changes 6
    The "New" Dimension 6
    Multilines, an architect's best friend 8
    Named levels 9
    Shared Cells 10
    Reference files: a few enhancements 10
    Rendering, a new set of tools 11
    New B-spline Commands 11
Summary 11

Chapter 2

The "New" Graphic environment
WHAT ARE ALL THOSE THINGS ANYWAY?

Windows 13
    Stacking Windows 14
The Command Window 15
    File Menu 16
    Edit 17
    Element menu 18

    Settings menu 19
    View menu 20
    Palette menu 22
    User menu 23
    Help menu 24
Tool Palettes 24
    Tools 26
Dynamic Panning 26
Next up... 27

Chapter 3

The "New" Elements

Dimensioning 29
    Setting up the dimensions 29
Saving dimension setups 33
    The Dimensioning Tools 35
    Associated Dimensions 36
    Modifying dimensions 37
    The "new" dimensioning tools 39
    Freezing and Thawing for 3.x compatibility 41
Multilines 41
    Setting up the multi-line 42
    Multi-Line Styles 43
    Changing a wall 43
Element Selection Tool 44
3D, a new set of tools 45
    Place Slab 47
    B-Spline Tools 48
Rendering 49
    Lighting and Rendering 49
    Controlling the lighting 50
    Ambient Lighting 50
    Flashbulb 51
    Solar Light 51
    Using light source cells 52
    The Shading options 53
    Constant shading 53
    Smooth Shading 54
    Phong Shading 55

Finally... 56

Chapter 4

The Tool Palettes
FINDING YOUR WAY AROUND

    PLACE LINE 59
    CONSTRUCT BISECTOR 60
    CONSTRUCT BISECTOR LINE 61
    CONSTRUCT PERPENDICULAR TO 62
    CONSTRUCT PERPENDICULAR FROM 63
    CONSTRUCT TANGENT BETWEEN 64
    CONSTRUCT LINE AA 1 65
    CONSTRUCT LINE AA 3 66
    CONSTRUCT LINE AA 2 67
    CONSTRUCT LINE AA 4 68
    CONSTRUCT TANGENT PERPENDICULAR 69
    CONSTRUCT LINE MINIMUM 70
    CONSTRUCT TANGENT FROM 71
    CONSTRUCT TANGENT TO 72
    PLACE LINE ANGLE 73
    PLACE CIRCLE CENTER 75
    PLACE CIRCLE DIAMETER 76
    PLACE CIRCLE EDGE 77
    PLACE CIRCLE RADIUS 78
    Tool Settings window 79
    CONSTRUCT TANGENT CIRCLE 3 80
    CONSTRUCT TANGENT CIRCLE 1 81
    PLACE CIRCLE ISOMETRIC 82
    PLACE ELLIPSE CENTER 83
    PLACE ELLIPSE EDGE 84
    PLACE ARC CENTER 86
    PLACE ARC EDGE 87
    PLACE ARC RADIUS 88
    CONSTRUCT TANGENT ARC 3 89
    MODIFY ARC ANGLE 90
    MODIFY ARC AXIS 91
    PLACE BLOCK 93
    PLACE BLOCK ROTATED 94
    PLACE SHAPE 95
    PLACE SHAPE ORTHOGONAL 96

PLACE ISOMETRIC BLOCK	97
PLACE POLY INSCRIBED	98
PLACE POLYGON CIRCUMSCRIBED	99
PLACE POLYGON EDGE	100
PLACE LSTRING	101
PLACE CURVE	102
PLACE MLINE	103
PLACE TEXT	106
Text Editor window	107
PLACE TEXT ABOVE	108
PLACE TEXT ON	109
PLACE TEXT ALONG	110
PLACE TEXT FITTED	111
PLACE NOTE	112
COPY	115
ROTATE (ORIGINAL)	116
COPY PARALLEL DISTANCE	117
COPY PARALLEL KEYIN	118
SPIN ORIGINAL (or copy)	119
ARRAY RECTANGULAR	120
ARRAY POLAR	121
MODIFY	123
INSERT VERTEX	124
DELETE VERTEX	125
DELETE PARTIAL	126
EXTEND LINE	127
EXTEND LINE KEYIN	128
EXTEND LINE INTERSECTION	129
EXTEND LINE 2	130
FILLET MODIFY	132
FILLET NOMODIFY	133
FILLET SINGLE	134
CHAMFER	135
MIRROR VERT/HORIZONTAL/LINE	137
MEASURE DISTANCE ALONG	138
MEASURE DISTANCE POINTS	139
MEASURE DISTANCE PERPENDICULAR	140
MEASURE DISTANCE MINIMUM	141
MEASURE RADIUS	142
MEASURE ANGLE	143
MEASURE AREA POINTS	144
MEASURE AREA ELEMENT	145
PLACE FENCE BLOCK	147

PLACE FENCE SHAPE 148
MODIFY FENCE 149
FENCE COPY 150
FENCE ROTATE COPY 151
FENCE ARRAY POLAR 152
FENCE MIRROR COPY VERTICAL 153
FENCE DROP 154

Chapter 5

Common Functions

MENUS, ATTRIBUTES AND OTHER COMMONLY USED ITEMS

The Sidebar menu 155
Design Options 157
Working Units 158
. . . . 158
. . . . 158
Cells 160
Text Attributes 162
Patterning 163
Element Symbology Control 164
Plotting 168

Chapter 6

Utilities

MCE, MICROSTATION MANAGER AND OTHER TOOLS.

The "Other" Utilities 173
   Importing and Exporting files 173
Exporting other types of data 177
   2D to 3D and back 177
   The Renderman connection 177
   The RIB file 178
Visible Edges 178
A Word about the future 179

Chapter 7

Updating your system to 4.0
SOME CONSIDERATIONS

Running concurrent versions (PC version only) 181
Developing Dimensioning Control 181
Hardware Considerations 182
    Video Considerations 182

Chapter 8

MicroStation For The IGDS User
A STUBBORN IGDS USER'S GUIDE TO MICROSTATION

Similarities Between MicroStation and IGDS 183
Where MicroStation and IGDS Differ 184
    The Databuttons 185
    View Manipulation Commands 185
    Panning around 187
    Design Options Tutorial, or the Lack Thereof 188
    The Plotting Menu 188
In A Word... 189

Index

Begins on page 191

Introduction

Introduction
HOW TO READ THIS BOOK

The *MicroStation 4.X Delta Book* is a quick, graphically oriented review of the differences between earlier versions of MicroStation and the new MicroStation 4.0. It is intended to introduce MicroStation and IGDS users to MicroStation 4.0's new graphical interface and many of the new functions that have been added to MicroStation.

Numerous command illustrations show how things used to be done and how they are down now. There are also chapters devoted to major areas of change with discussions of many of MicroStation's new functions.

How This Book Is Organized

MicroStation 4.X Delta Book is set up so it can be quickly read to gain an understanding of the fundamental differences between MicroStation 4.0 and earlier MicroStation versions. It is also indexed so that it works as a reference to guide you through the transition.

Early chapters review what's new in MicroStation 4.0, describe the user interface and discuss the new element types. The heart of the book is Chapter 4 which has a number of element placement command illustrations that show how a command worked on earlier versions of MicroStation and how the command works now. This section also includes a number of illustrations for commands that are new to version 4.0.

Chapter 5 reviews other commands and functions that are not strictly element placement ones such as level control, element symbology, menus, attributes and plotting. Chapter 6 reviews utilities such as MicroStation Manager, DXF and the Renderman interface. Chapter 7 discusses strategies for upgrading to 4.0, and Chapter 8 is a short IGDS user's guide to MicroStation 4.0.

The index not only includes entries for commands and concepts, it includes an extensive index of illustrations.

Command Illustrations

To help orient you to MicroStation, a "standard" illustration of individual commands was developed. This appears in two forms in this book. The one for pre-4.0 versions of MicroStation shows the commont ways of selecting the command on all three MicroStation platforms (MicroStation Mac, MicroStation 32 and sidebar menus typically used with MicroStation PC).

With MicroStation 4.0, the user interface is the same on all platforms. The new command illustrations reflect this by showing only the icon on the tool palette that activates the command. The following diagram shows a typical command illustration for MicroStation 4.0.

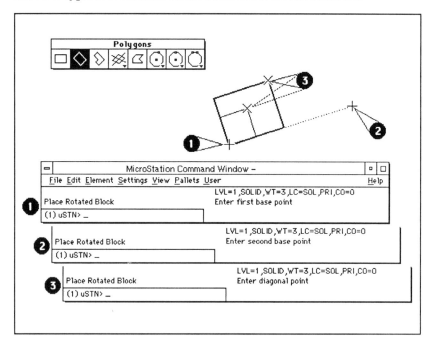

In each illustration there are three distinct regions.

Along the top are tools for command selection. The command window appears at the bottom of the illustration. As the command unfolds each numbered window shows what MicroStation says.

In the center of the figure the actual operation of the command is shown with the numbered "meatballs" associating which action goes with which command window prompt.

In some instances an additional picture element (marked with an arrow) shows how the command concludes.

What Is Not Covered in this Book

This book is intended to be a quick review and a helpful guide to the differences between the new and old MicroStations. It is not designed to provide detailed instruction on new or previously existing commands. There are a number of other books that provide a wealth of detailed information about MicroStation 4.0.

INSIDE MicroStation by Frank Conforti and the *INSIDE MicroStation Companion Workbook* by Mike Ward provide excellent introductory level refernce and instructional material for MicroStation 4.0. The *MicroStation Productivity Book* by Steinbock, Kincaid and Malm is a good advanced user's book that has been fully revised and expanded to cover MicroStation 4.0.

In addition to these books, there are a number of others listed in the back of this book that cover such topics as working in 3D, user commands and MDL. When you need more information about MicroStation 4.0, consider purchasing these books.

MicroStation™ 4.X Delta Book

A Quick Guide For Upgrading to 4.X

By Frank Conforti

Chapter 1

Welcome to MicroStation 4.0

A TOOL FOR THE NINETIES

Welcome to the MicroStation 4.0 Delta book. During the course of reading this entertaining yet thoughtful book, it is hoped you will become familiar with the most advanced software package to leave Intergraph. That's right, MicroStation 4.0 represents more than a "minor" revision upgrade; it is a whole new product.

Previous versions of MicroStation took their cue from Intergraph's other powerhouse product, IGDS (short for Interactive Graphic Design System, but then you knew that). The backbone of Intergraph's Vax based products, this venerable workhorse of a CAD program has been the standard by which most CAD packages have been compared. In fact MicroStation 3.x (x for X marks the system, 3.3 for PC, 3.4 for Interpro, 3.5 for Mac) still does not have all of the functionality of IGDS (such as polygon clipping of reference files).

MicroStation 4.0 is a milestone in Intergraph history. It is the first version of MicroStation to radically depart from the IGDS look-and-feel. It is also the first to employ new design tools for which there are no parallels in IGDS. Finally 4.0 has the distinction of being the first CAD package to embrace an industry standard graphic user interface across all hardware platforms!

The "New" Look

Sporting its 90's look, the first thing you notice about 4.0 is the graphics. Taking its cue from Apple's Macintosh and Microsoft's Windows 3.0 product, MicroStation 4.0 now has a very graphical user interface that follows OSF/Motif GUI (acronym for Graphical User Interface) guidelines. This look is not unique to MicroStation. In fact once you have mastered 4.0 you will be ready to operate other Motif based products.

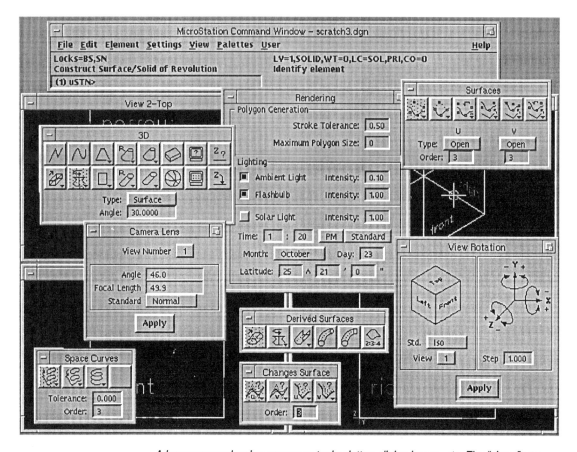

A busy screen showing numerous tool palettes, dialog boxes, etc. The "views" can just be seen in the background.

Intergraph has done more than just dress MicroStation up in its new GUI. As a further sign of its commitment to the Motif standard, they now bundle Looking Glass, a Motif desktop "shell" (written by the Visix Corporation), with every Interpro workstation. This means that users of their Workstation products will not be able to "tell where the operating system ends and the CAD begins". Furthermore, all of Intergraph's application software has been written to run under Looking Glass as well.

So, the new look and feel of MicroStation 4.0 really is part of a larger effort to enhance Intergraph's overall system approach to design. What this means to Mr. (or Ms.) CAD USER is no matter which computer platform MicroStation runs on, its operation will be exactly the same. Furthermore, all application software written in MDL,

MicroStation's newest programming environment, are fully portable between all platforms!

MicroStation Development Language

Not to belittle some of the other new features of MicroStation 4.0 there is one new addition that will have more impact in the long run than even the "new" graphics. That is MicroStation Development Language (MDL for short). Essentially an implementation of industry standard 'C' language this new programming environment is exciting for a number of reasons. First, its portability (as just mentioned). If you write a program in MDL on a PC and then port it over (read: copy) to an Interpro workstation, you will be able to run this program as is. No modifications, no compilations, no nothing.

MDL also has another endearing feature, its ability to tie in tightly to the MicroStation operations. When you first fire up MicroStation 4.0 you may be presented with the following display:

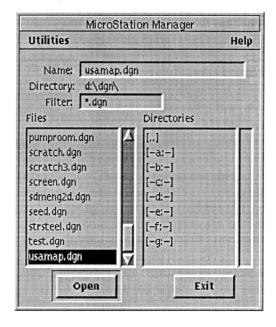

MicroStation Manager is an MDL program.

Unbeknownst to you this "manager" is nothing more than a very sophisticated MDL program. You have pulldown menus, click on buttons and all of the "other" trappings of the Motif/MicroStation program. In fact many of the functions and tools found in MicroStation are MDL applications.

With time many third party software developers and Intergraph itself will offer extremely powerful, discipline specific applications to run within MicroStation. Users of these products may not even know where MicroStation ends and these third party packages begin.

Back to Basics: the Changes

Other major enhancements to MicroStation involve new and changed element types. MicroStation uses a specific set of "primitive" element types from which you build your design. Two new element types have been added: the Dimension element and the Multiline element.

The "New" Dimension

As most users of Intergraph know, the 3.x dimension is nothing more than a set of lines, cells (maybe) and text loosely associated with the graphic group feature to document measurements. Changes to a dimension quite often resulted in awkward editing of these various elements, or just delete them and try again. Under 4.0 this has changed dramatically.

MicroStation now sports a new primitive element, the dimension. Among its new capability is one of special interest, its associativity. At your discretion you can now "attach" a dimension to an object within a design, whether a line, arc or whatever. When you modify that object (such as move, or stretch it), the dimension will automatically update.

Back to Basics: the Changes   1-7

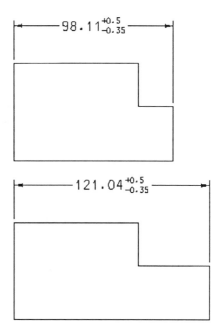

Associative dimensions.

Other enhancements to dimensioning includes full editability and the incorporation of dimensioning standards called Styles. By selecting an appropriate style of dimension you can make your dimensions conform to a chosen industry standard.

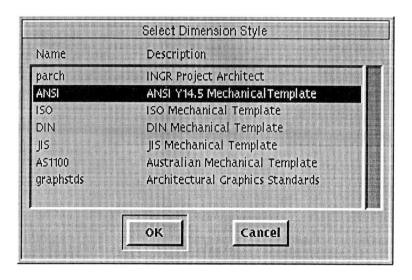

The dialog box from which you can choose which dimension style to use. Additionally you can create your own style and store it with these industry giants.

Dimensioning is covered in a later chapter.

Multilines, an architect's best friend

The other major new element type is the Multiline. This simple but very powerful element consists of two or more parallel linestrings that act as one element. You control the appearance of each linestring as well as the quantity of them. Up to 16 parallel linestrings make up the multiline. What makes this element different than copying a normal linestring parallel is in its behavior.

Using a series of Multiline Joint tools you can modify the appearance of a multiline and, for instance, cut a hole in it for placing a door. What is different here is the two halves of the multline "wall" are still one element. If, for instance, you wish to move a vertex of the wall all you have to do is use the Modify tool and move it as if it were a linestring. All of the parallel lines would move accordingly, including any "holes" you may have cut in the multiline.

The other powerful feature of the multiline is the "style" file. Just as with the dimension element a style library can be created in which you store the most common multiline styles by name. For instance,

Intergraph supplies styles for many of the common wall types found on a floor plan.

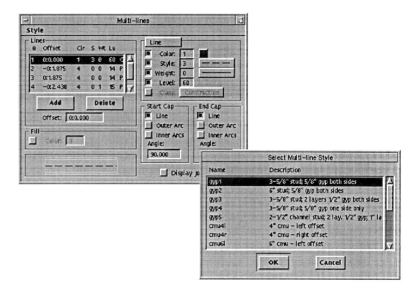

An example of a multiline style as called up by name. Note the description and the appearance of the graphic as a wall sheathed in gypsum board.

Named levels

This concept of naming "things" has also been extended to levels. You can now assign a descriptive "name" to individual levels, assign a selection of level names to a group name and use these names as you would level numbers. For instance, you can turn a group level off, say "building (levels 1,10,11,14)" by keying in OF=building. MicroStation treats this as if you had keyed in the level numbers one by one. The setup of these levels is controlled by the Level Names window.

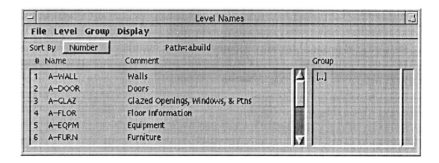

The Level Names window where you edit the level names.

Shared Cells

A new option has been added to the cell functions. Using a toggle lock (SET SHARECELL or Use Shared Cell button) you can control the basic method of how a cell is placed in your design file. With SHARECELL OFF, cells are placed as always, each occurrence a copy of the cell's elements. With SHARECELL ON it's a different story.

Instead of copying a cell's elements over and over again as you place a cell, the "shared" cell only needs to record the location of the cell. All of the element's of that cell are stored once in the design file out of harm's way. This means that a particularly complicated cell, such as a pattern cell (concrete comes to mind), does NOT take up TONS of disk space!

In addition, when you update a shared cell (REPLACE CELL) all occurrences of that cell reflect the change. One step.

Reference files: a few enhancements

Finally catching up to IGDS, MicroStation 4.0 now handles polygon clipping of reference file bounds. Taking this one step further you can also "blank" out areas of a reference file. Called a mask, this feature also can be any shape.

Rendering, a new set of tools

In the 3D arena MicroStation has increased the number of rendering options. There are now eight distinct rendering options available. From simple hidden line removal through the advanced Phong and Gourand shading techniques, there is even a 3D stereo option! Of course you will need your 3D glasses for this last one to work.

Additionally, you can set the drawing perspective with each view. Normally MicroStation displays everything in parallel view. Real world, however, is subject to the vanishing horizon (everything recedes to the horizon). For more realistic appearance of a rendering, MicroStation now supports a perspective selection process analogous to a camera. You select an appropriate "lens" (which controls the breadth of the view and the foreshortening aspect to the horizon) and position it in relation to the object you are rendering, as if you were setting up to take a photograph.

To add to this realism you can also load a raster image as a "background" to bring in details to a rendering (Clouds in the sky, for instance).

New B-spline Commands

A number of advanced b-spline creation and editing tools have also been added to MicroStation's toolkit. They allow you to more easily construct complex #D b-spline surfaces. These are briefly reviewed later in the book.

Summary

This quick overview of what's new with MicroStation 4.0 is neither detailed or complete. Almost every aspect of this software package has had the once over. This has led to little changes here and there to enhance the operation of the overall product.

During the whole makeover process nothing was lost: no capability; no command. In fact you can still use your "old" sidebar menu, digitizer command menu, or even key-in command names with MicroStation 4.0. The intent for the rest of this book is to illustrate the major differences between earlier versions and MicroStation 4.0 as well as show you the "new" tools and how they work.

Chapter 2

The "New" Graphic environment

WHAT ARE ALL THOSE THINGS ANYWAY?

With a little understanding of some of the changes the time has come to take a tour of the new graphics environment.

Windows

Every time you select a dialog box or open a view or even Open a Design file, you are in fact activating a window in the GUI. All of these various windows have a number of common features. They are: a border (sometimes resizeable), a title area, and control buttons.

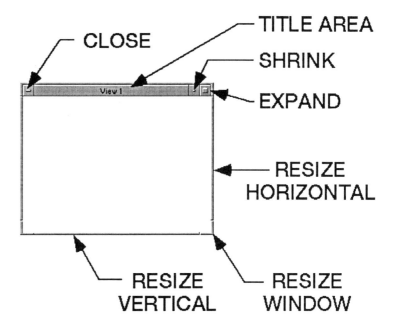

A "typical" view with all of its parts labelled. NOTE: The close button is also called the Cancel button.

The window title bar contains a number of features, all for controlling the overall disposition of the window. The buttons in the corners may: expand/contract the view, push it under other views, or cancel it altogether. This latter operation is access via the "dash" button in the upper left corner of the view. Selecting this button and dragging your cursor down to the Close selection will close the view. A short cut is to double click on this button.

The Cancel button (labeled the Close button in the previous diagram) activates this menu.

To move a window you datapoint in the title area and drag the window to where you wish it to appear. On most dialog boxes you can also move them by selecting the window border in a similar fashion. But don't try this with a view window. Being a resizable window, dragging view's border will result in the view expanding. Resizable windows are designated by the "cut" marks near each corner. If you click and drag one of the corners, the result is a resizing of both the vertical and horizontal surface.

Stacking Windows

By now you have seen the way MicroStation allows you to overlap windows and "stack" them. Any time you select a window by its title area it pops to the top of the stack and is active. Selecting a different window pops it to the top, etc. You can "sink" a view by selecting Sink from the menu opened with the Cancel button.

The Command Window

One common window you have already witnessed is the Command Window. This window works exactly the same as 3.x's command area. The active command name or "tool" appears in the command area. So do the tool prompts as well as error messages. In fact, this command window should reassure you somewhat; the order and location of the various fields (CF,PR,ER,IF, etc.) are all the same as they are in 3.x.

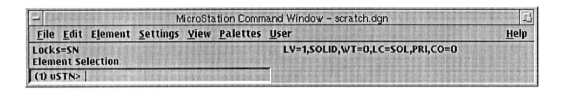

The command window has only one new feature: the pulldown menubar.

There is, however, one new feature: the pulldown menubar. This is one of the main methods of communicating with MicroStation 4.0. By selecting and "pulling down" on each word you are presented with a number of selections. Most selections, in turn lead to a dialog box or tool palette.

File Menu

Grouped by common function, each pulldown menu services a specific area of MicroStation. File, for instance manages the opening, creation, and attaching of design files and cell libraries. Additionally this is where the most common of the file manipulation selections are found. Importing DXF and Text is done from here as well as exporting DXF and RIB files. The obligatory FILE DESIGN is also found here now labelled Save Settings. Compress design file is also on the File menu.

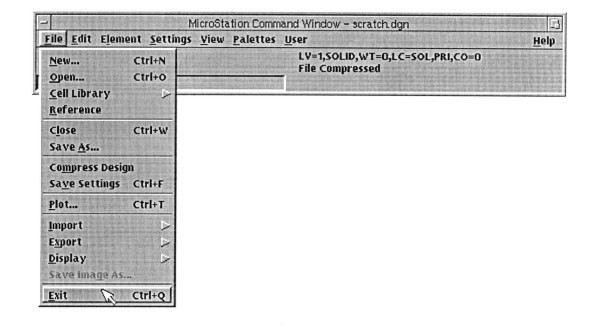

The EXIT command is also found on the File menu.

Edit

The Edit menu is where the undo/redo functions reside as well as a new capability: orphan cell creation (now known as Grouping).

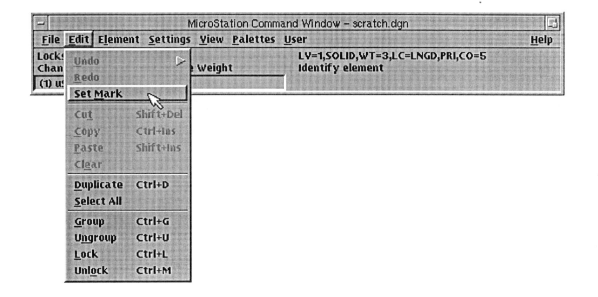

The Edit pulldown menu.

Element menu

From the Element menu you can interactively select your colors, line styles and weight from a popup menu. In addition there are a number of "specialty" element settings available from here: Dimension setup, multilines, b-splines, and text.

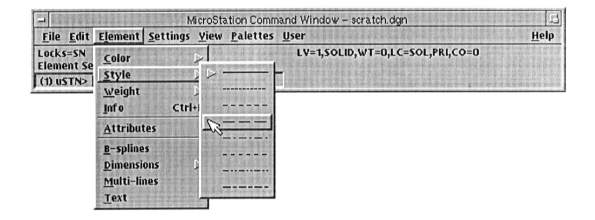

Dragging your cursor over the styles example and releasing the datapoint selects that style.

Settings menu

The Settings menu can be likened to the design options tutorial. Here, however, you activate the specific function you wish to "set". Selecting the Grid selection, for instance, brings up the Grid settings window where you set the various grid attributes. Under 3.x you accomplished this using key-ins exclusively. You can still do this via the command window, which in many cases is probably faster. However, with some settings windows (like Locks) it may be easier to leave them open and pushed to the back, available all of the time.

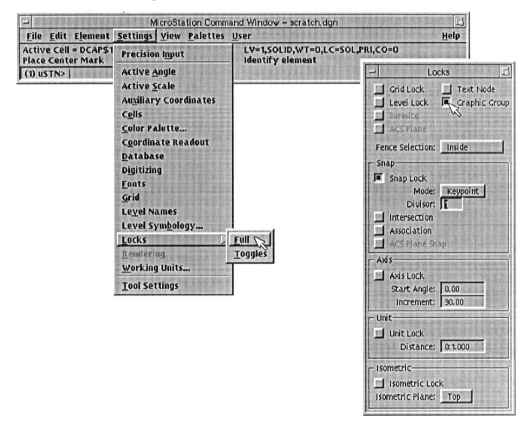

The Settings menu.

View menu

This is the menu that controls all of the various features of the working views. All of the traditional commands are available from here: Update, View ON/OFF, fit, window area, and the other standard view commands. Additionally, there are a number of new commands and functions affecting your views. The Attributes selection brings up a settings window where you set the display ON/OFF parameters such as weight display, grid display, etc.

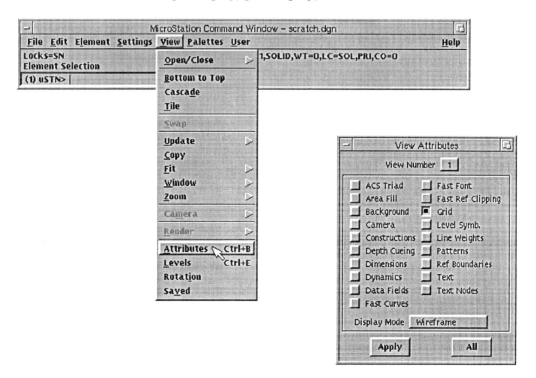

Additionally, the view menu is where the Rendering and Camera features are found. You can also access many of the more common view commands such as update, window area, zoom in/out and fit from the Main tool palette.

Palette menu

As the name suggests, this is the menu from which you select the various tool palettes you use to access the numerous design tools. The most important one of these is the Main tool palette selection. The "Grand Central Station" of graphic tools, this palette is on the top of the list, so to speak.

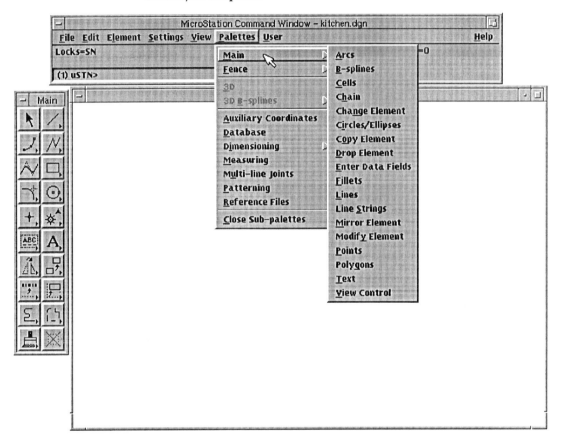

The Main palette.

User menu

This short menu is where you control the overall or "global" functions of MicroStation. From here you select Preferences, which brings up the Preferences settings window. Of most importance is the Compatibility value. Set to 4.0, you can utilize all of the new features of MicroStation 4.0. Set to 3.x, many of the new element types are converted to their constituent components (read: dimensions=lines and text).

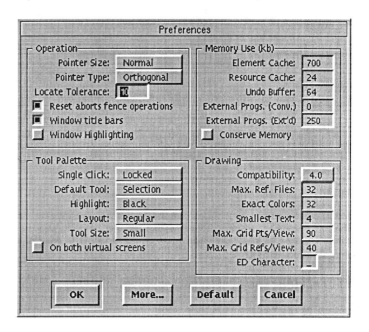

Help menu

A good example of an MDL program in action is Intergraph Help section. A context sensitive help, this on-line information source gives you descriptions of commands as you select them. The Command Browser lets you "build" your command as a series of selections of keywords.

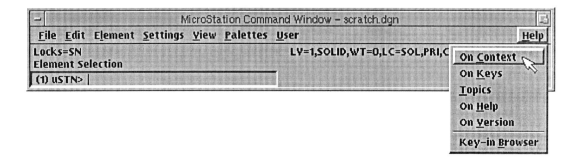

Tool Palettes

Analogous to the sidebar menu of 3.x fame, the Tool palettes represent a new method of selecting "commands". Each tool on a palette is represented by a unique icon. After a while these icons will become as familiar to you as the command locations were on the USTN sidebar menu.

Floating as a separate window on the screen, a tool palette may spawn sub-palettes. This is similar in fashion to the sidebar first column selection changing the second column listing. The difference here is the second column can be "torn off" which results in a separate floating sub-palette.

The Main Tool palette invoked from the Palette pulldown menu actually controls most of the common graphic tools. Grouped by element type or function, almost all of the Main Palette tools pop up a sub palette. Selecting the line tool will result in a pop up sub-palette containing all of the other line construction tools.

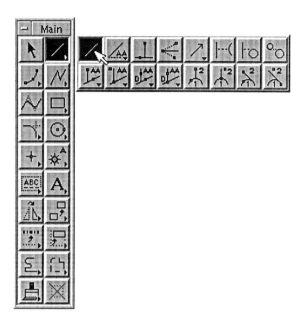

The Place Line tool really hides a complete set of line construction tools.

When you tear off a sub-palette an additional function will become apparent. When you select a command that requires additional parameters (such as a circle's radius) a pop down field appears under the sub-palette. You select these fields as needed and adjust the value displayed. This replaces the traditional MicroStation 3.x function of prompting you for the missing information.

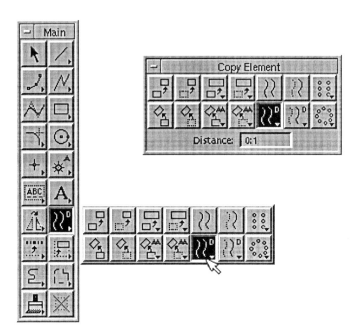

An example of a sub-palette before and after it is torn off. In this case there is a popdown data field associated with the command chosen (COPY PARALLEL KEYIN).

Tools

Popdown fields are always indicated by an arrow in the lower right hand corner of the tool icon. This popdown only appears if the subpalette has been torn off of the main palette. As an aid the most recently selected tool on a subpalette will appear on the Main palette and the location associated with that palette. As you drag your cursor across tool icons (browsing) you will notice that the name of the tool appears in the command window.

Dynamic Panning

A new and exciting feature added to 4.0 is a new method for scrolling around the drawing. Rather, it's an old capability associated with the old Interact and 68000 graphic workstations. By holding the shift key down while datapoint and dragging in a view, you pan the view around the design plane. This can be done at any time in any command.

This is done without special hardware, although if you have certain video cards (8514's, TIGA equipped cards) MicroStation will take advantage of their on-board processor. However, this dynamic panning works on all video cards. The speed at which the panning takes place is controlled by how far you move the cursor from the initial point at which you datapointed and the processing speed of your computer. In any case this is one of the more important capabilities shoehorned into MicroStation 4.0.

Next up...

That concludes an overview of the new graphics environment and its trappings. Now let's review some of MicroStation's new elements.

Chapter 3

The "New" Elements

This chapter deals with the specifics of the new element types and "new" features.

Dimensioning

MicroStation's "other" major area of change is in dimensioning. Going from the "sticks and text" method of dimensioning to a "smart" element approach does involve major changes in the user of dimensions. First of all the traditional tutorials used to set up the dimension elements have been replaced by a number of new settings boxes.

Setting up the dimensions

Dimension setups are accessed via the Element pulldown menu:

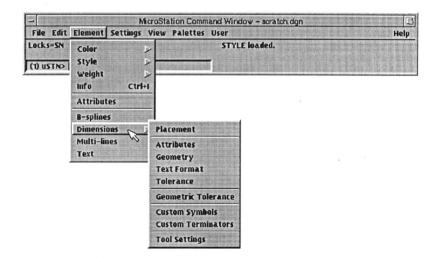

The pulldown menu for selecting the various dimension settings.

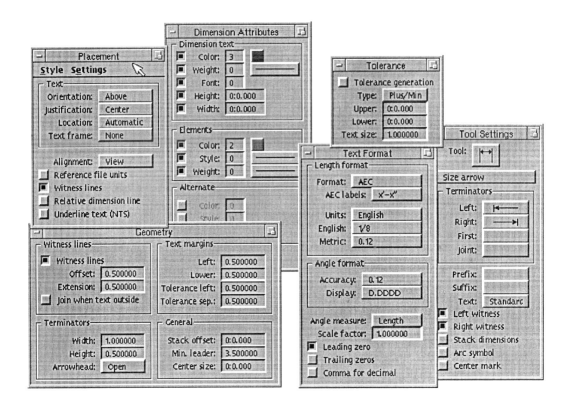

All of the various settings boxes needed to control the dimensioning process.

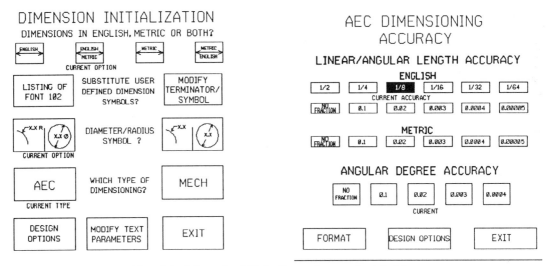

The original dimension initialization and accuracies tutorials...

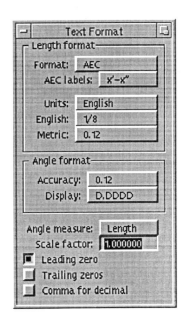

...and the closest equivalent in dimension settings windows: the Text Format window.

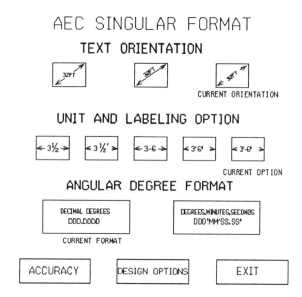

The "old" tutorial for setting up the placement of the text and labelling options...

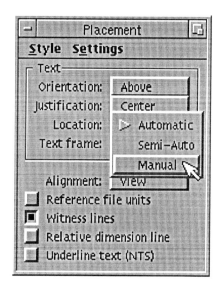

...and the "new" settings window for controlling these actions.

NOTE: The labelling of the dimension itself is set in the Text Format settings window along with the angle dimension format.

Probably the largest hurdle to overcome when switching to MicroStation 4.0 dimensioning is the totally new organization of the various control parameters. Now, instead of being organized by discipline (mechanical vs. architectural) all parameters are grouped by function. All parameters affecting text are grouped together. All dimension attributes (as in element attributes) are together, as are the text placement controls.

The whole idea behind the dimensioning organization is to supply you with those tools that change frequently in one settings window, while other less frequently changed parameters have their own dialog box.

In reality the various settings windows are a combination of the "traditional" tutorial parameters and the sidebar (or command menu) dimensioning control selections.

Saving dimension setups

Probably the saving grace in all of this seemingly complex set of menus is the ability to save the dimension setups in a separate file. Called Styles, this new function stores most of the dimension setups for later recall. In fact, Intergraph supplies an extensive set of industry standard styles encompassing most of the major dimensioning standards.

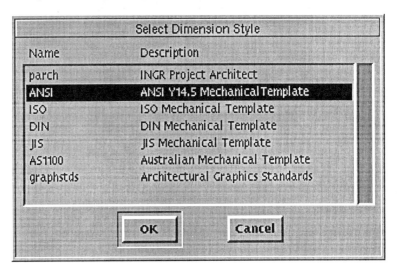

The Select Dimension Style is accessed via the Style pulldown menu on the dimension Placement window.

3-34 MicroStation 4.X Delta Book

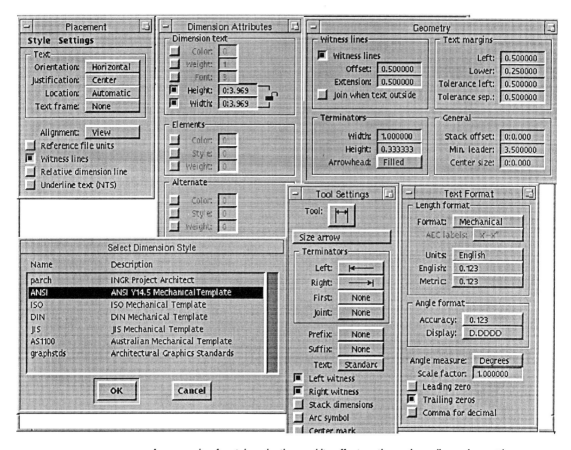

An example of a style selection and its effect on the various dimension settings windows.

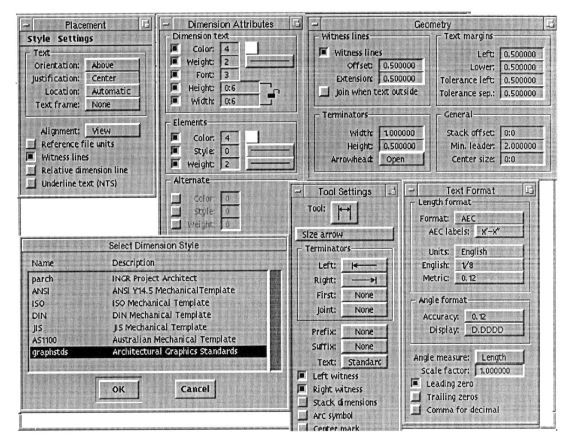

Another example of a style selection. Compare this illustration to the previous one to note the changes.

The Dimensioning Tools

Accessed via the Palettes pulldown menu, the various dimensions are grouped into four sub-palettes.

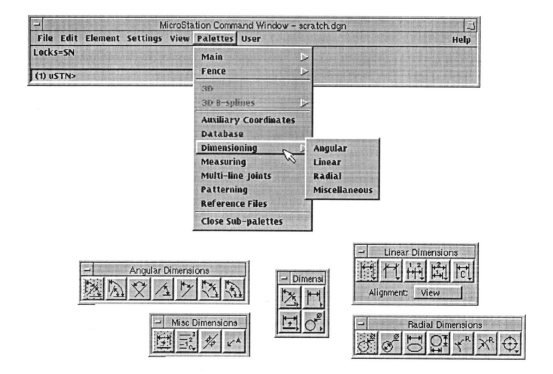

All the dimensioning tools.

Each palette supports a specific type of dimension: Linear, Radial, Angular and Miscellaneous. Most of the tools shown are already familiar to you and work in much the same way as under 3.x. There are, however, a couple of "minor" differences from before.

As you performed the dimension command MicroStation would build your dimension "on the fly". If it was not to your liking you deleted it (or used UNDO) to remove. Under 4.0 a dimension is not "set" until you hit two resets, similar to the way a linestring is created. The first reset still allows you to "turn a corner" when dimensioning.

Associated Dimensions

The other major change to the dimension tools is the associated dimensions. By turning on the Snap Association lock (from the Settings/Locks pulldown menu) all dimensions placed with the tentative

Saving dimension setups   3-37

point will be associated to the element selected. This is an important new feature!

The Association lock is found on the Settings/Locks/Full pulldown settings window..

When an associated dimension is placed, its endpoints do not actually have a coordinate value. Instead the element (or intersection) supplies this coordinate information to the dimension. Thus, if you modify the element the dimension changes as well.

Modifying dimensions

Editing existing dimensions has never been easier. Because a dimension is now a "primitive" element, the normal modify element tools works on them as well. For instance, to split a dimension you use the Insert Vertex tool:

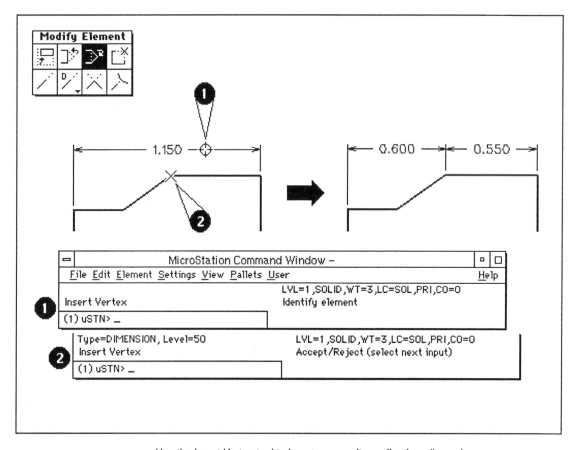

Use the Insert Vertex tool to insert a new witness line in a dimension.

To change a non-associated witness line location, use the Modify Element tool:

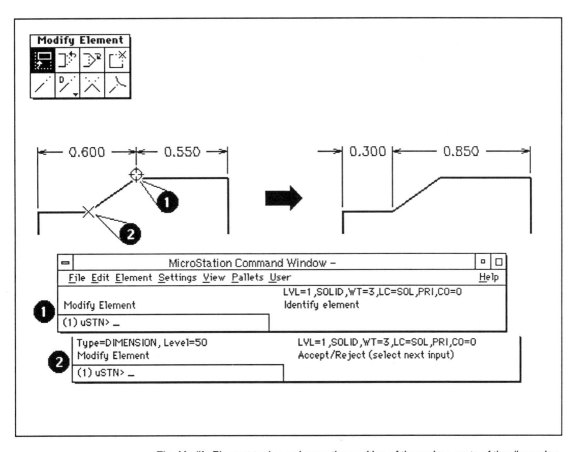

The Modify Element tool can change the position of the various parts of the dimension.

The "new" dimensioning tools

In addition to all of the commands found in 3.x, a number of new dimensioning tools have been added:

Dimension Size (Custom):	Allows you to place a linear dimension in line with an isometric drawing.
Dimension Element:	Dimensions a selected element in one step.
Dimension Diameter (Extended Leader):	Draws a dimension line across the circle with arrowheads and a leader line to the text.
Dimension Radius (Extended Leader):	Draws a line with arrowheads from the center of arc/circle to dimension text.

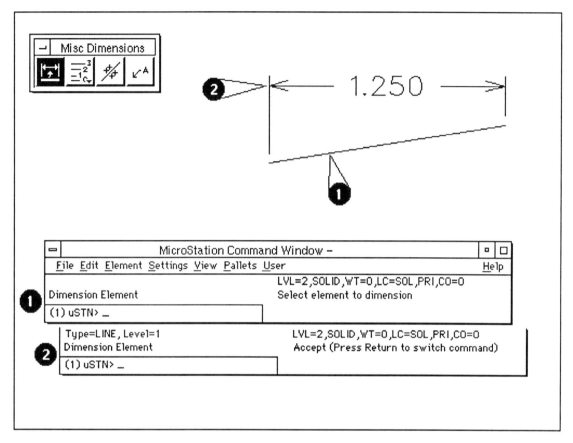

The Dimension Element tool. Hitting consecutive carriage returns prior to the second datapoint will cycle through the various options (Size Arrow, Size Stroke, Label Line).

All of these new features of the dimensioning package are only available as part of 4.0. If you must maintain compatibility with 3.x (or IGDS) then many of these features will not be available to you. Compatibility is controlled via the User Preferences dialog box selected from the User pulldown menu.

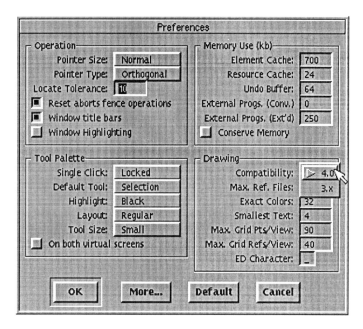

Selecting the Compatibility popup menu gives you the choice of 4.0 with all of its enhancements to the dimensioning routines, or 3.x with its individual "sticks and text". Alternately you can key in SET COMPATIBILITY ON/OFF.

Freezing and Thawing for 3.x compatibility

There is a way to maintain the advanced features of 4.0 if you need to temporarily transfer a file to 3.x. Using the Freeze Element (key-in: FREEZE), you in essence create a duplicate of the selected dimension rendered in lines and text. The "real" dimension is sequestered away. In this way you can provide a 3.x user the ability to view or even modify a 4.0 dimension.

However, upon returning the design file to 4.0, any changes done under 3.x will be lost when you THAW the dimension. Thaw Element is the compliment to Freeze. The lines and text associated with the 3.x version will be lost forever, thus the reason for the inability to pass changes from 3.x to 4.0.

Multilines

The "other" major new element type introduced with MicroStation 4.0 is the Multiline. Acting like a super linestring, this element is destined to be used by the architectural design community. In essence the multiline is a primitive element that consists of a number of parallel lines that act like one element. In addition, the ends of the multiline can be terminated with either a line, arc or none.

What makes the multiline so powerful, however, is its editability. Using a series of "cutter" tools, you can modify a multiline to provide openings for doors, windows and other egress. All of these modifications do NOT change the fact that the multiline is one continuous element. This is in contrast to using the Partial Delete tool on an element. For this reason MicroStation has created a separate palette just for the Multiline Joint tools.

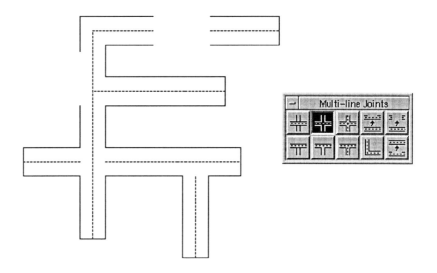

Multiline joint tools.

Setting up the multi-line

Multi-lines are controlled by the Mutli-line settings window. Accessed from the Element pulldown menu, this dynamic window is where the action is. The part of the Multi-lines dialog box that has the most bearing on its use is the Lines section. Here you set the number and spacing of the lines that will make up your multiline. Highlighting one of the lines and selecting the Delete button eliminates the definition of that line. Add is simple: you select the Add button and the offset, color, style and weight displayed in the labelled fields will be applied to the new line added.

The Offset field just mentioned is where you set the distance between your datapoints and where the parallel line will be drawn. If you set no offset (i.e. zero) then the resulting line will be drawn similar to the linestring. If on the other hand you key in a distance either positive or negative, your multiline will be drawn offset from your datapoint.

Multi-Line Styles

As with the dimension elements, Multi-lines supports the use of a style file. Used to store common sets of multi-lines, it is available from the Style pulldown menu on the Multi-line settings window. Intergraph delivers a good example of how this style file works.

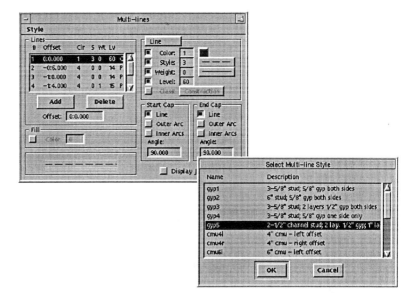

Each multi-line style can be fairly complicated. In this example the gyp5 results in a representation of a stud wall with gypsum board covering.

Changing a wall

Multi-lines are subject to the same element modification commands as linestrings. Additionally you can change from one style to another using the CHANGE MLINE key-in command. In this way you can have a change of heart during a design session and not have to re-enter the multi-line.

Element Selection Tool

Not truly a new element type but equally as important is the introduction of a new element manipulation "tool". Called the element selection tool, this device turns the whole CAD process around. Actually what it does is reverse the normal method of selecting a command, selecting elements for the command to operate, and a final datapoint for acceptance. The *selection set* will be familiar to any Macintosh user who has used practically any of the Macintosh graphics programs currently available.

The arrow tool located in the upper left corner of the main tool pallet is the secret to this "new" manipulation method. You select one or more elements using the SELECT ELEMENT command (the arrow tool) and then select the particular element manipulation command desired. Highlighted elements are noted by the appearance of small filled boxes called "handles". The result will be immediate and usually will not require a final acceptance datapoint. Selection sets work with all of the element manipulation commands.

Another feature of the Element Selection command is its ability to act like an element modifier command. The "handles" that are displayed when an element is selected are the keys to this tool's operation. By datapointing on a highlighted element's handles, you can change the location of the endpoints of a line, the diameter of a circle, or the corner of a shape. If you datapoint and drag somewhere along an element's length, the result is similar to the move command. This is as close as you can get to a modeless operation in MicroStation.

You can select more than one operation by datapointing and dragging a diagonal across the elements desired. Their handles will appear noting their selection. You can add additional elements or deselect previous elements by datapointing while holding down the shift key.

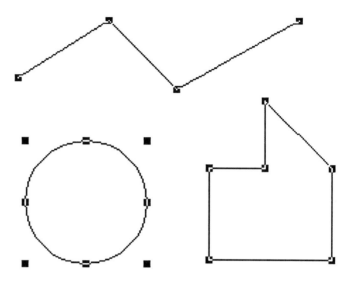

Some elements showing their selection set "handles".

3D, a new set of tools

If all this new element types wasn't enough, Intergaph has enhanced the 3d features of MicroStation considerably. New b-spline tools have been added along with excellent view manipulation tools designed to make it easy for you to find your way around.

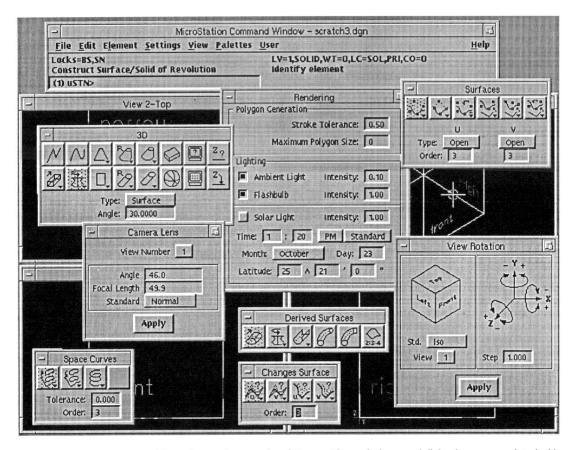

The various palettes, sub-palettes, settings windows and dialog boxes associated with 3D.

With MicroStation 4.0, a new tool has been provided to make it easier to visualize the twisting and turning that inevitably occurs during a 3D design session. The View Rotation dialog box is accessed from the View pulldown menu. Consisting of a animated cube representing the six "standard" views, this tool allows you to specify rotations along the three standard axes by simply clicking on the stick figure representing these axes.

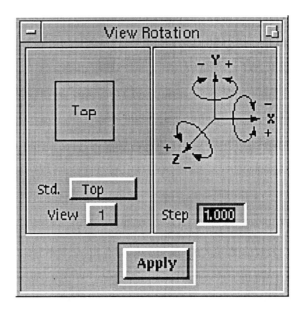

The View Rotation tool presents you with a very effective method of rotating any view.

Place Slab

Recognizing that many element projections performed in a design session result in orthogonal shapes (i.e. cubes and boxes) MicroStation 4.0 provides the PLACE SLAB command. By defining two corners of a rectangle and a depth using an adjacent view, you can create such boxes in a hurry. The advantage with this command is you don't have to deal with active scales, active angles or even active depths!

Three simple datapoints and you've got a box!

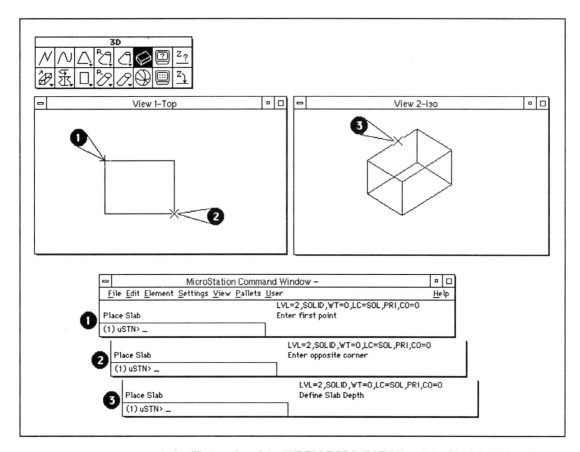

A simplified version of the SURFACE PROJECTION tool, the PLACE SLAB makes it easy to create perfect boxes everytime.

B-Spline Tools

MicroStation 4.0 adds a number of powerful new b-spline surface tools for creating 3d curved objects. The following diagram shows a boat hull created with the Construct B-spline Surface by Cross-Section tool

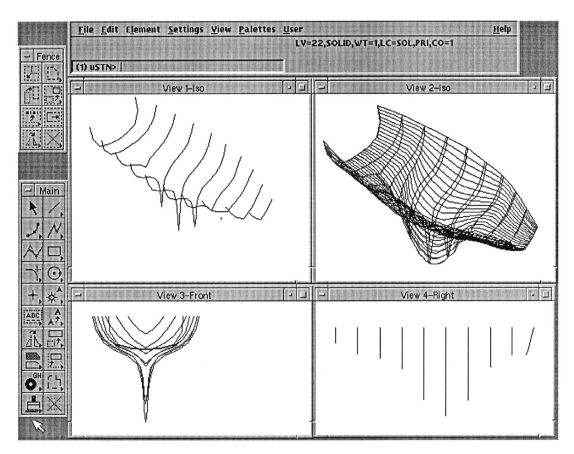

3D boat hull drawn with the Construct B-spline Surface by Cross-Section tool.

Rendering

Along with all of the new 3D elements tools and view control tools, the rendering capability has been greatly enhanced. Now there are a number of ways to render a 3D design from crude and quick to complex and slow. But first, you need to light up those images.

Lighting and Rendering

Rendering requires you to set up a light source for illuminating your model. This is done via the Rendering settings window (from the View pulldown menu). The three shading options perform not only a hidden line function but also calculate the brightness of the various surfaces within your design. How bright an object appears takes into consideration a number of variables. The most important aspect is the object's angle to the light source.

When you select any of the shading options with the default lighting source, MicroStation sets the light source as that of the viewer. This would be as if you were shining a flashlight at the view. And just as with a flashlight, the image does not quite look "real". Think of how things look in the headlights of a car. To better control the look of a shaded image MicroStation provides a number of lighting options.

Controlling the lighting

Selecting the Rendering settings box from the Settings pulldown menu, you are presented with a number of lighting options.

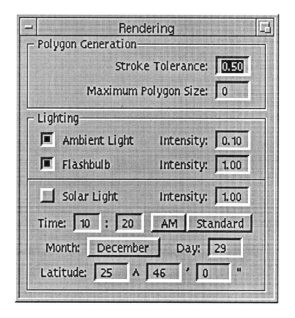

The lighting settings box for selecting the general lighting parameters.

You have three choices: Ambient, Flashbulb and Solar. Although you can turn all three on, this diminishes the effect of the individual lighting options.

Ambient Lighting

An artificial light source, this option controls the uniform illumination surrounding your rendering. Similar to general office lighting, ambient light source is good for highlighting model details that would otherwise be too dark to see. The default value of 0.10 (i.e. 10%) is a good value. Anything over 0.4 will wash out the colors of your rendering.

Flashbulb

The default lighting source, this is the light that is "shined" on the elements from your viewing perspective. This can be adjusted from 0.0 to 1.0 (100%). Lower values darkens the colors used to shade the surfaces.

Solar Light

A very interesting option, the Solar Light source simulates the direction and intensity of light as it would appear coming from the sun. By setting the Latitude and the Date-and-time, you can approximate how your project would appear on a given day. This is most appropriate for architectural renderings but can be applied to any project.

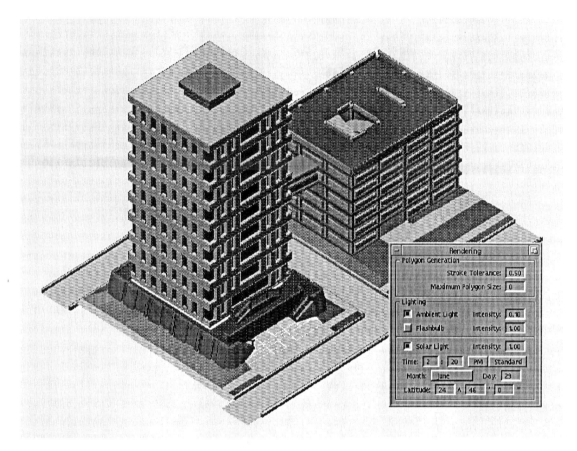

An example of a Solar Lighted project

Using light source cells

The standard three light sources given so far address most lighting situations. There are, however, times when you may want to specify a lighting source more precisely. Lamps along a street may need to be simulated to show their effect on an architectural project. A proposed consumer product may be shown in a better "light" by selecting light sources similar to how a photographer sets up a "shoot".

MicroStation provides this with the LIGHTING.CEL library. Containing a number of specialized cells, this cell library offers you the control over the lighting necessary to meet the most demanding lighting conditions. There are three cells provided:

   PNTLT - point light source (360 radiation)

DISTLT - distant light source (more diffuse)

SPOTLT - spot light (very directional)

Each cell in turn contains enter data fields where you enter the Red, Green, and Blue intensities thus controlling the color of the light generated.

The Shading options

All of this talk of lighting leads us to the actual shading operations. The three choices in shaded rendering are: CONSTANT, SMOOTH and PHONG.

Constant shading

By far the simplest form of shading, this option breaks curved surfaces into a series of facets. Each facet is in turn colored with a solid color calculated with a flat face at an angle to the various (or one) light source. The results is a rough computer looking image.

The roughest of the three shading techniques CONSTANT SHADING works best with non curved objects. It is good for quick results.

Smooth Shading

The next best shading method is SMOOTH. Again using a faceted approach to breaking down a curved surface, the difference is in how these individual facets are treated. The shading algorithm includes an operation that "averages" the shading from one facet to adjacent ones. This method produces a smoother appearance by sort of blurring the image.

The SMOOTH SHADE option gives you a more realistic appearance. The look of the image still appears computer generated.

Phong Shading

The most computationally intensive of the three shading options, PHONG calculates the shading value AT EACH PIXEL LOCATION. This means if you are shading a VGA screen, the calculation for shading is done at each dot along each column, row by row. This would result in a calculation of 307,200 pixels. This can take some time! Now if you are using a higher resolution monitor of say 1,280 by 1,024 pixels, the number of calculations would jump to 1,310,720! Obviously this level of calculations means you should save this type of rendering for the final product, using the smooth and constant options for actual view rotation testing.

The PHONG shading option is the most computationally intensive of the renderings functions. The results are excellent if used with displays capable of displaying 256 colors or more.

Finally...

This chapter outlined some of the more important "new" developments in MicroStation 4.0. However, many of the changes found in 4.0 weren't big earth shaking CHANGES but rather small refinements to existing commands. The purpose of the next chapter is to show you some of those changes, comparing them to 3.x where appropriate.

Chapter 4

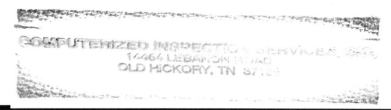

The Tool Palettes

FINDING YOUR WAY AROUND

In this chapter we are going to show you many of the graphic tools you have come to use and love. In many instances two illustrations are presented showing you where and how the command worked under 3.x and how it works under 4.0. Towards the end of the chapter only the 4.0 tool locations are shown.

Each section gives you the name of the tool palette or sub-palette and its cooresponding location on the Main tool palette. In the case of the Fence and Measuring palettes, their location is provided from the Palettes pulldown menu. Where there is a significant difference in the operation of a tool from its predecessor, a comment will be included in the caption.

What this chapter is not is a comprehensive listing of every tool known to man (or at least to MicroStation). Rather, it is supposed to show you examples of many of the most common tools you will be using on a day to day basis. Additionally, some element types are not covered here because there have been extensive enough changes in either the tools or the elements themselves to warrant coverage in their own chapter. Dimensions, for example, are covered in an different section.

Lines Sub-Palette tools

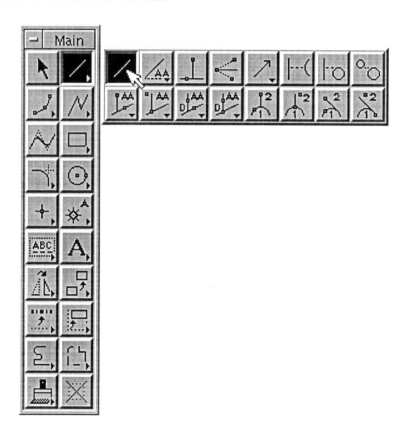

PLACE LINE

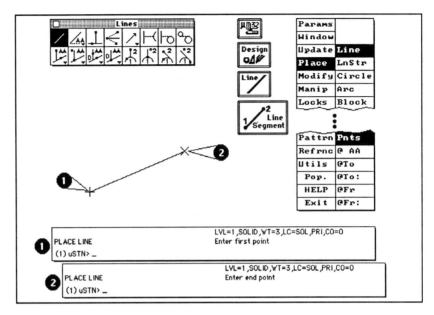

PLACE LINE command

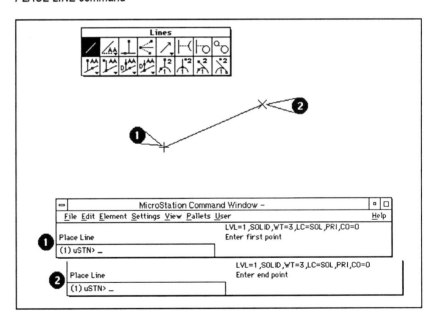

...and the Place Line tool

CONSTRUCT BISECTOR

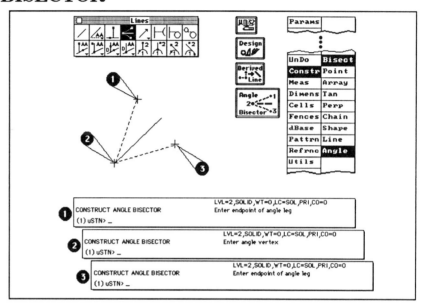

The CONSTRUCT BISECTOR command...

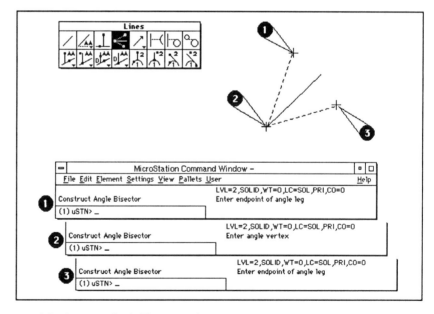

...and the Construct Angle Bisector tool.

CONSTRUCT BISECTOR LINE

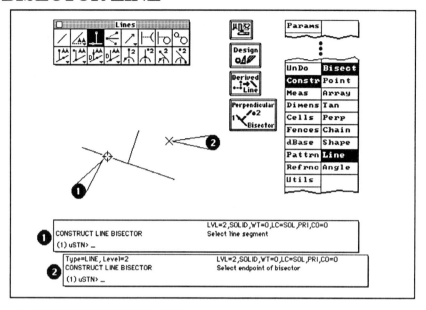

The CONSTRUCT BISECTOR LINE command...

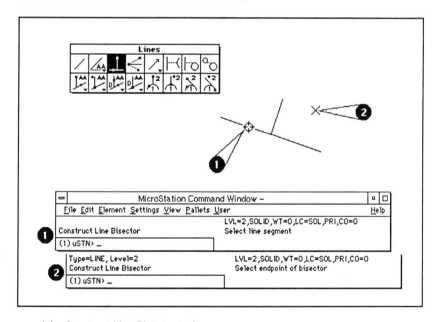

...and the Construct Line Bisector tool.

CONSTRUCT PERPENDICULAR TO

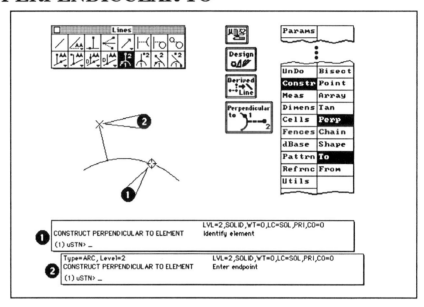

The CONSTRUCT PERPENDICULAR TO command...

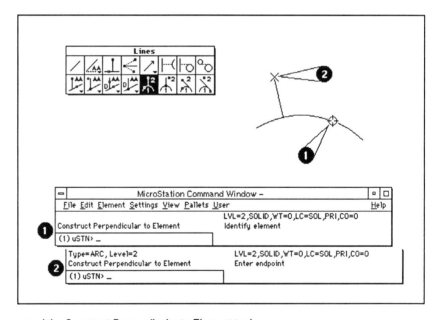

...and the Construct Perpendicular to Element tool.

CONSTRUCT PERPENDICULAR FROM

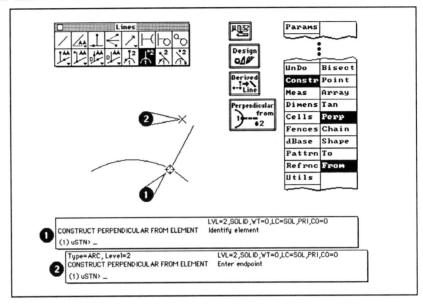

The CONSTRUCT PERPENDICULAR FROM command...

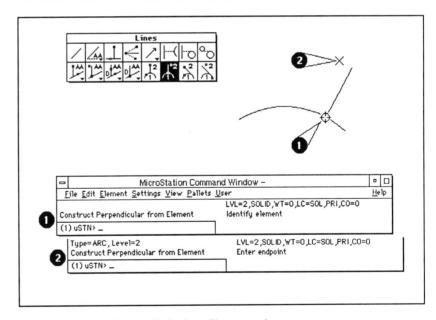

...and the Construct Perpendicular from Element tool.

CONSTRUCT TANGENT BETWEEN

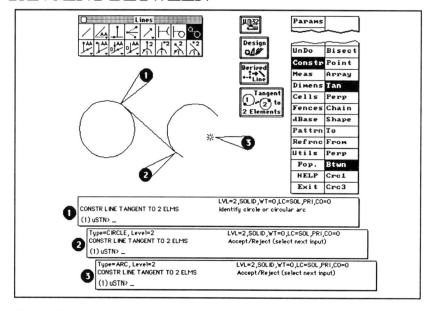

The CONSTRUCT TANGENT BETWEEN command...

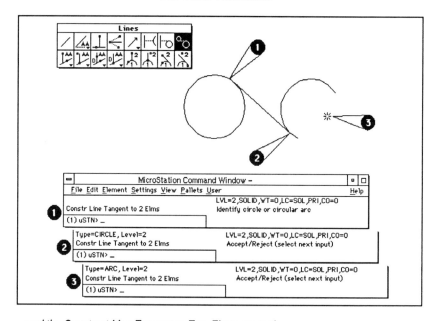

...and the Construct Line Tangent to Two Elements tool.

CONSTRUCT LINE AA 1

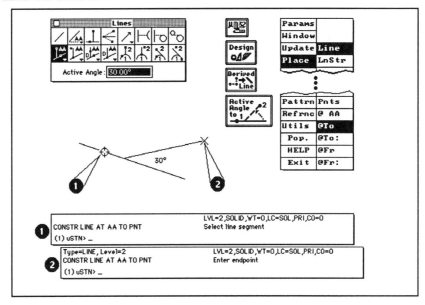

The CONSTRUCT LINE AA 1 command...

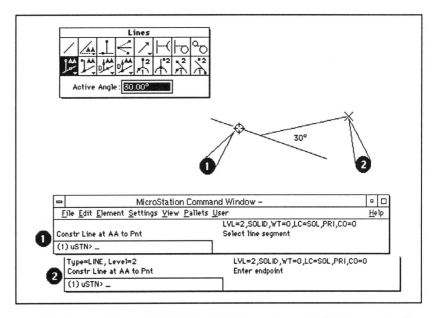

...and the Construct Line at AA To Point. Note the popdown data field for entering the active angle value.

CONSTRUCT LINE AA 3

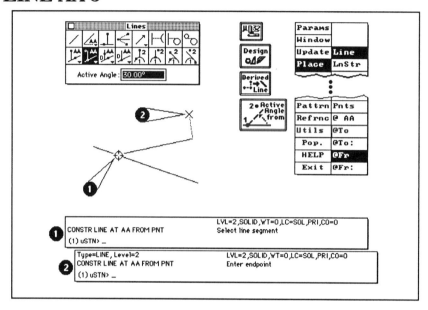

The CONSTRUCT LINE AA 3 command...

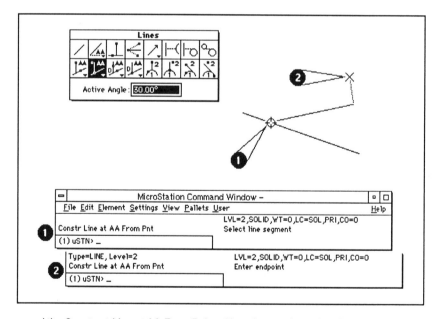

...and the Construct Line at AA From Point. Note the popdown data field for entering the active angle value.

CONSTRUCT LINE AA 2

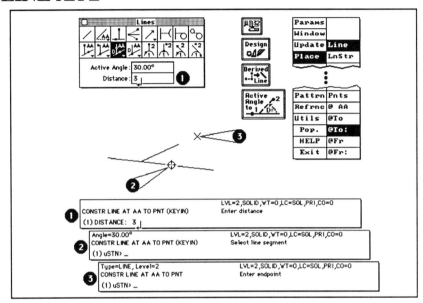

The CONSTRUCT LINE AA 2 command...

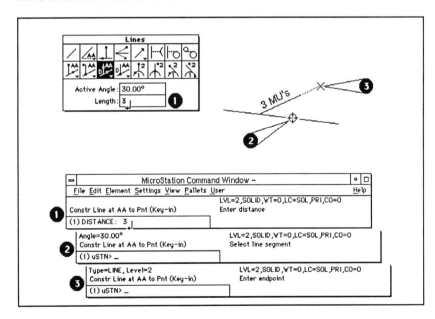

...and the Construct Line at AA to Point (Key-in). Note both the angle and the line length are entered through the popdown fields.

CONSTRUCT LINE AA 4

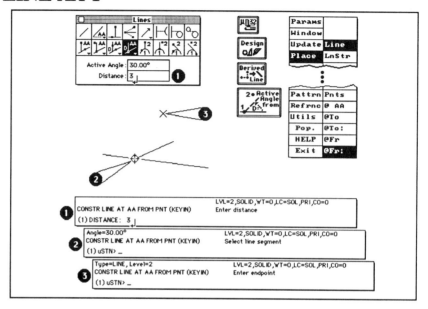

The CONSTRUCT LINE AA 4 command...

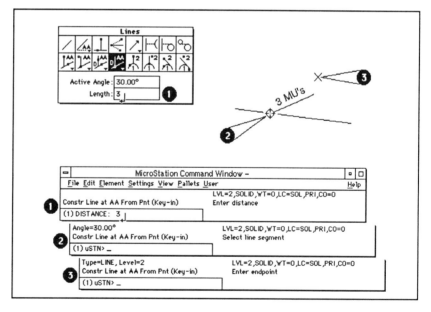

...and the Construct Line at AA From Point (Key-in). Note both the angle and the line length are entered through the popdown fields.

CONSTRUCT TANGENT PERPENDICULAR

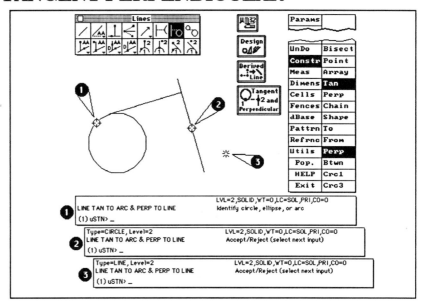

The CONSTRUCT TANGENT PERPENDICULAR command...

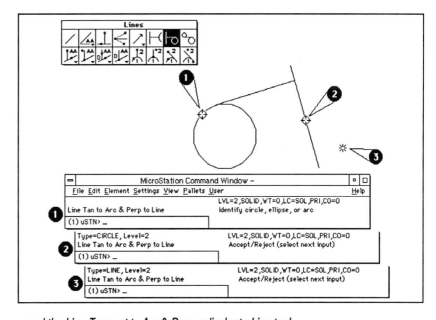

...and the Line Tangent to Arc & Perpendicular to Line tool.

CONSTRUCT LINE MINIMUM

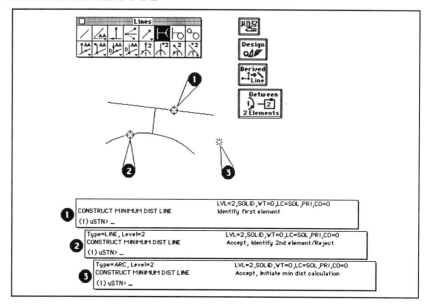

The CONSTRUCT LINE MINIMUM command...

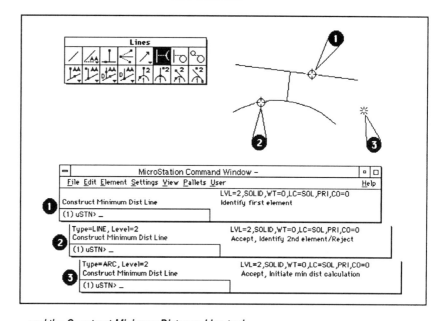

...and the Construct Minimum Distance Line tool.

CONSTRUCT TANGENT FROM

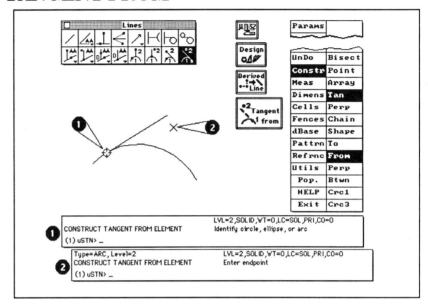

The CONSTRUCT TANGENT FROM command...

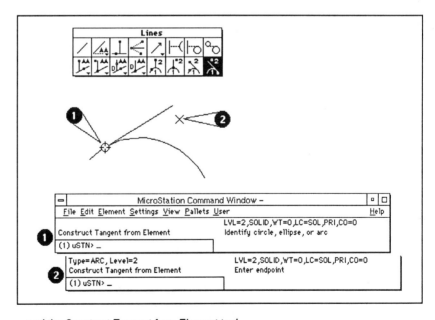

...and the Construct Tangent from Element tool.

CONSTRUCT TANGENT TO

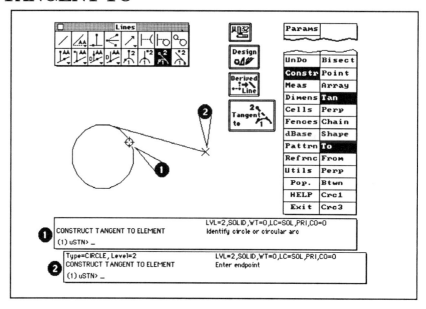

The CONSTRUCT TANGENT TO command...

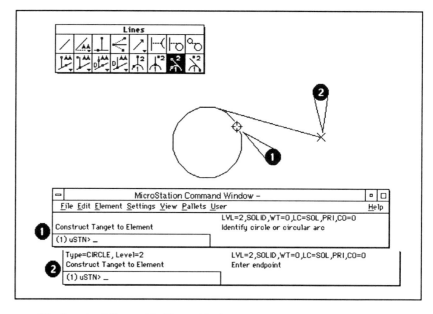

...and the Construct Tangent to Element tool.

PLACE LINE ANGLE

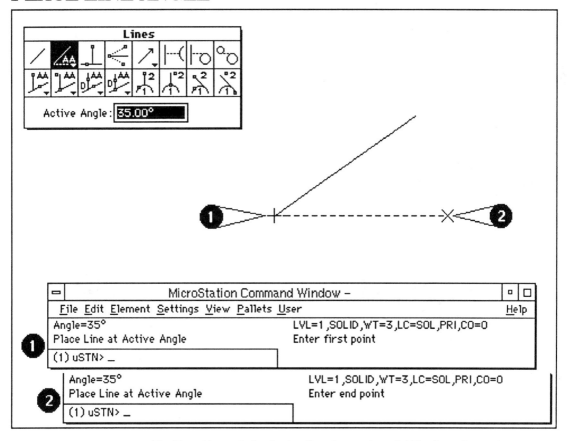

The Place Line at Active Angle. Note the popdown field for the active angle.

Circles/Ellipses sub-palette

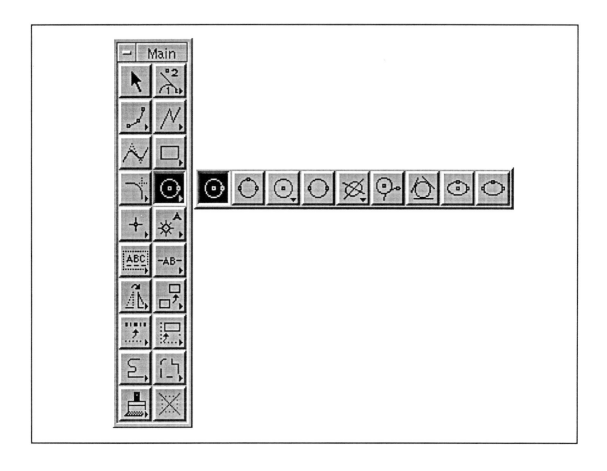

PLACE CIRCLE CENTER

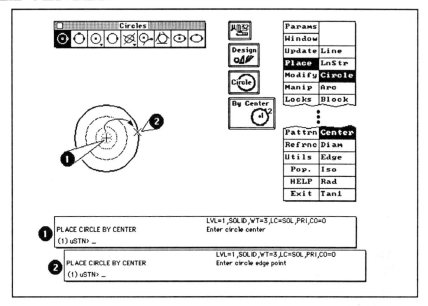

PLACE CIRCLE CENTER command...

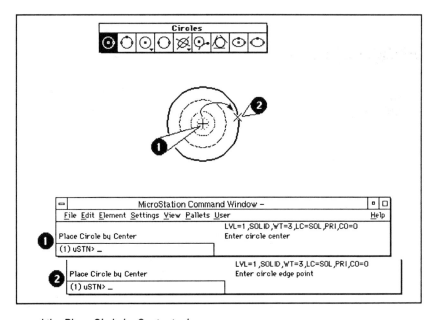

...and the Place Circle by Center tool.

PLACE CIRCLE DIAMETER

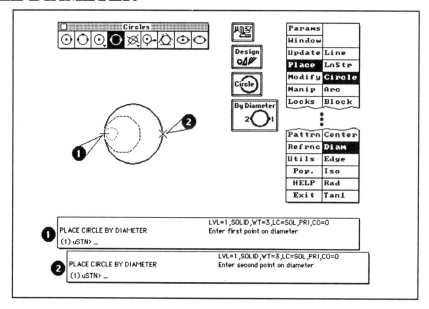

The PLACE CIRCLE DIAMETER command...

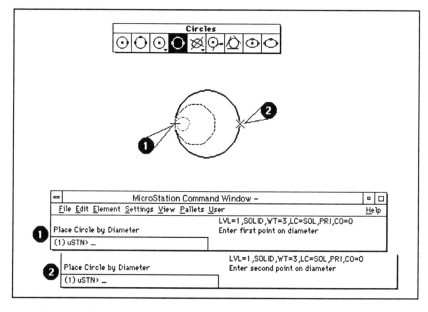

...and the Place Circle by Diameter tool.

PLACE CIRCLE EDGE

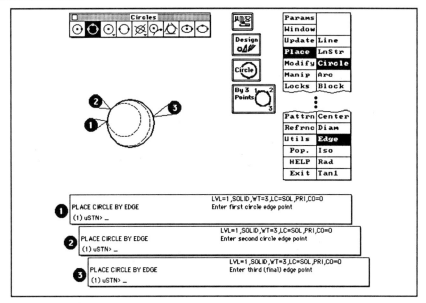

The PLACE CIRCLE EDGE command...

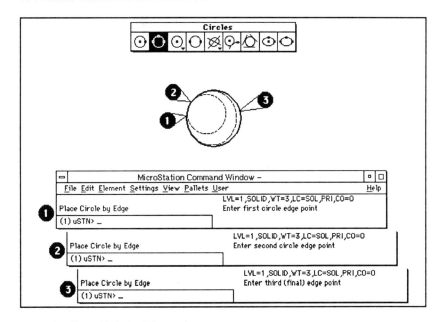

...and the Place Circle by Edge tool.

PLACE CIRCLE RADIUS

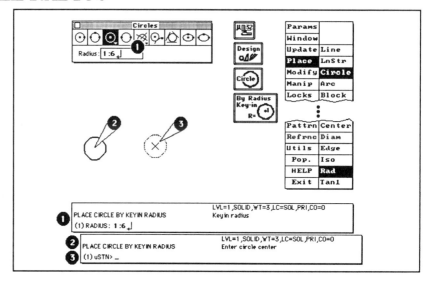

The PLACE CIRCLE RADIUS command...

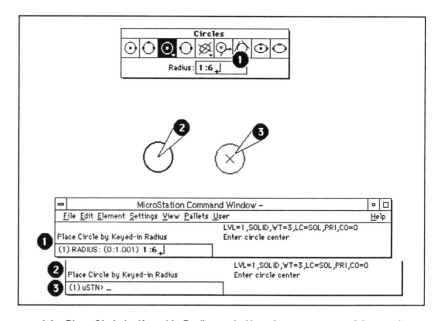

...and the Place Circle by Keyed-in Radius tool. Note the appearance of the popdown Radius field. This data field is only available when you activate the Circles sub-palette. Alternately you can activate the Tool Settings dialog box.

Tool Settings window

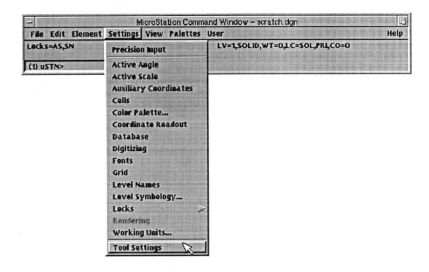

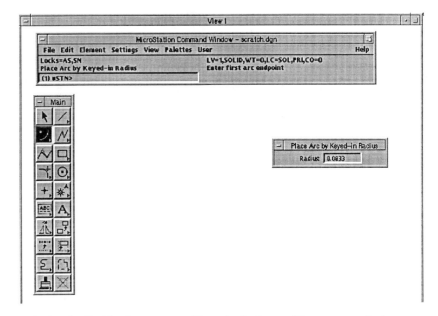

Selecting the Tool Settings command from the Settings pulldown menu activates a small window that displays the active tool and its popdown data fields.

CONSTRUCT TANGENT CIRCLE 3

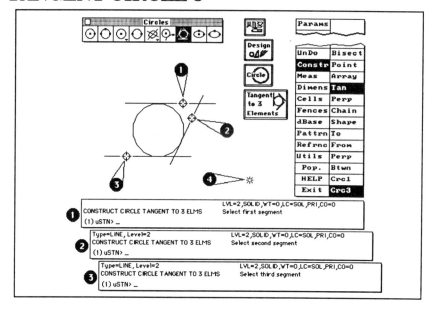

The CONSTRUCT TANGENT CIRCLE 3 command...

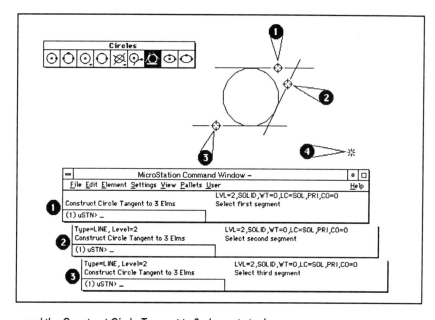

...and the Construct Circle Tangent to 3 elements tool.

CONSTRUCT TANGENT CIRCLE 1

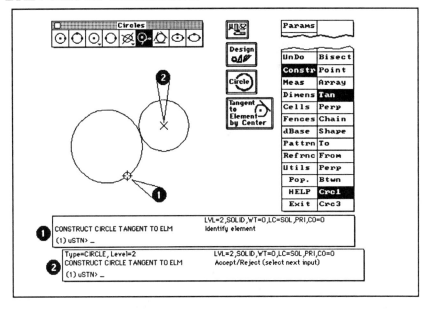

The CONSTRUCT TANGENT CIRCLE 1 command...

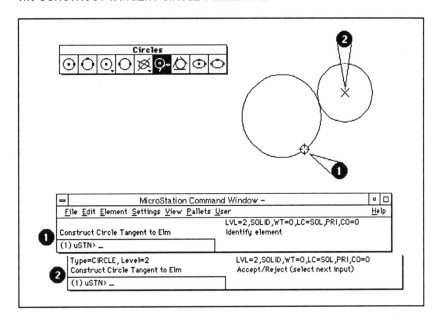

...and the Construct Circle Tangent to element tool.

PLACE CIRCLE ISOMETRIC

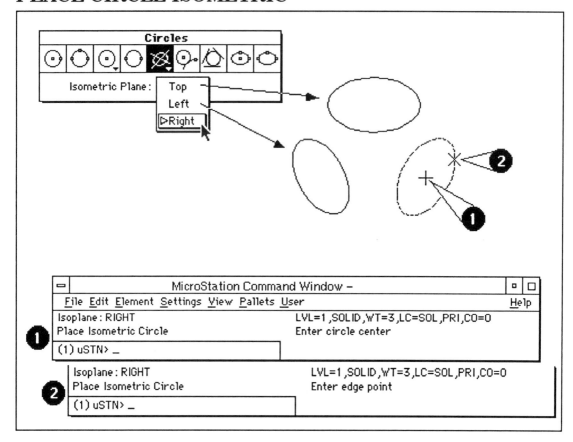

The Place Isometric Circle tool. Note the pop down data field for setting which "side" of the isometric view you wish the circle to be planar to. You select Top, Left or Right by clicking on the Isometric Plane field and selecting the appropriate "view".

PLACE ELLIPSE CENTER

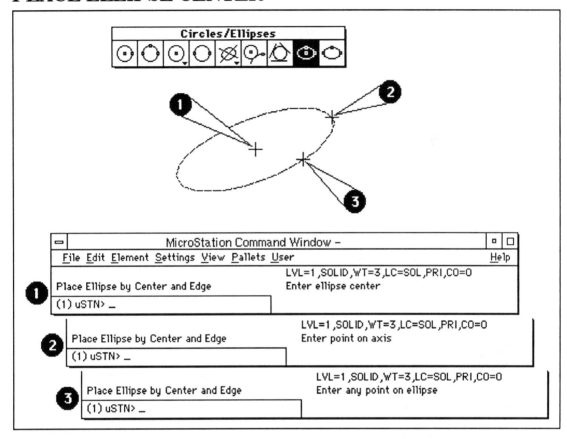

The Place Ellipse by Center and Edge tool.

PLACE ELLIPSE EDGE

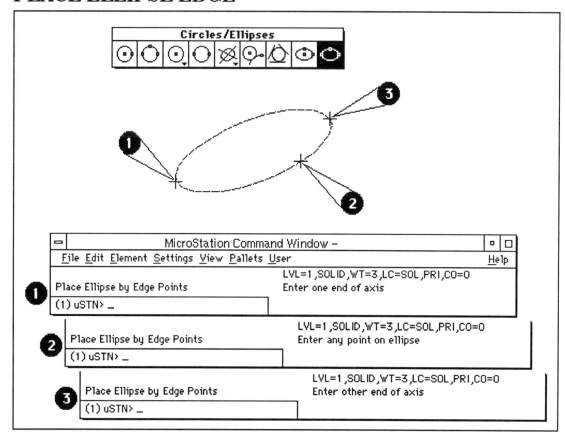

The Place Ellipse by Edge Points tool.

Arcs sub-palette tools

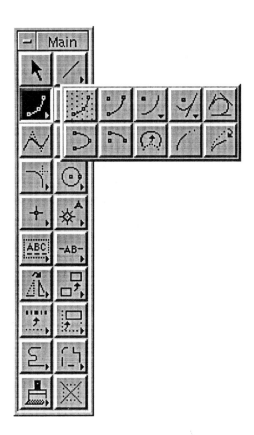

PLACE ARC CENTER

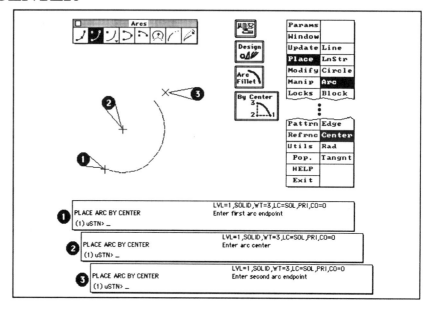

The PLACE ARC CENTER command...

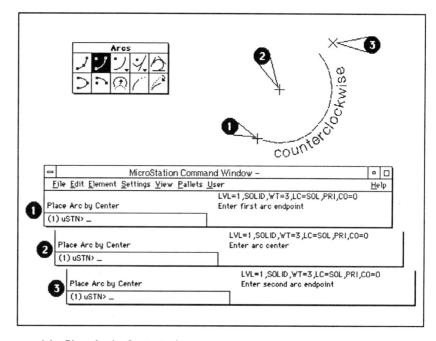

...and the Place Arc by Center tool.

PLACE ARC EDGE

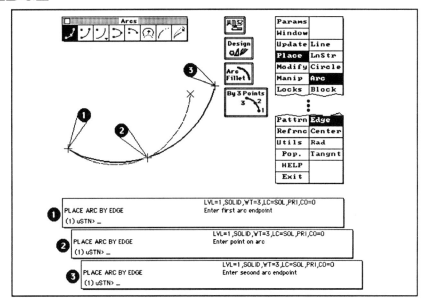

The PLACE ARC EDGE command...

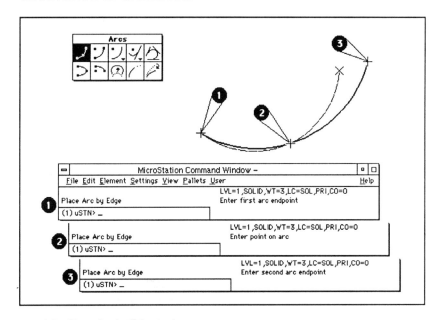

...and the Place Arc by Edge tool.

PLACE ARC RADIUS

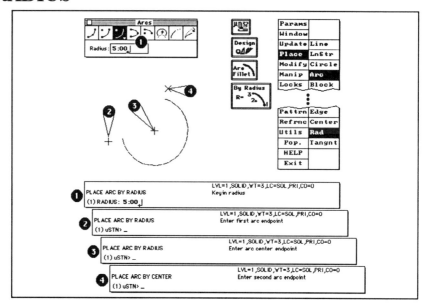

The PLACE ARC RADIUS command...

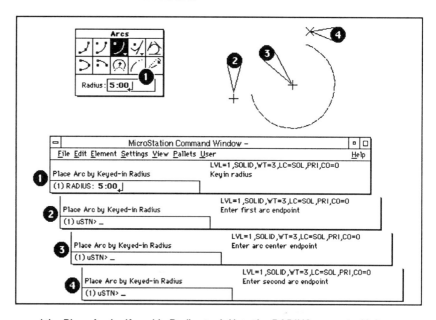

...and the Place Arc by Keyed-in Radius tool. Note the RADIUS: prompt with its "default" radius in parenthesis. This is displayed when you type the command name instead of selecting it from the tool palette.

CONSTRUCT TANGENT ARC 3

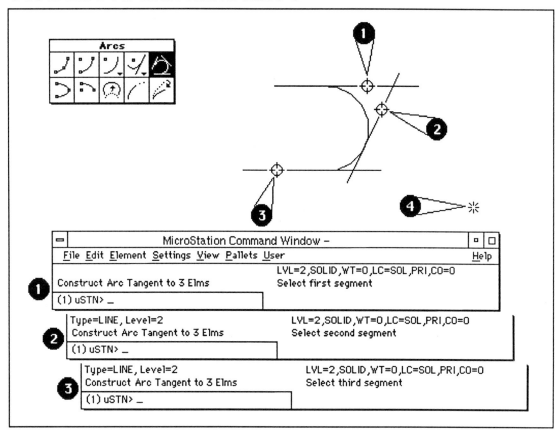

The Construct Arc Tangent to 3 Elements tool. Introduced with 4.0, this tool is similar to an IGDS command of the same name.

MODIFY ARC ANGLE

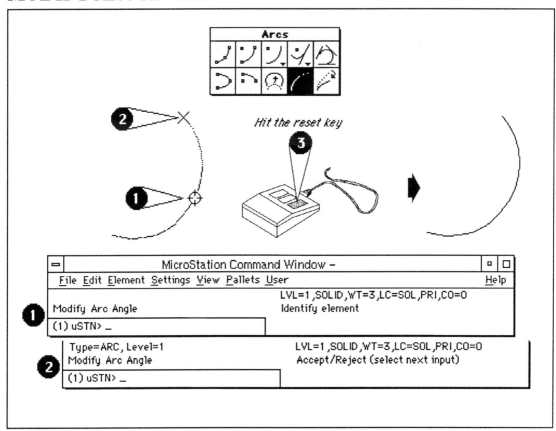

The Modify Arc Angle tool.

MODIFY ARC AXIS

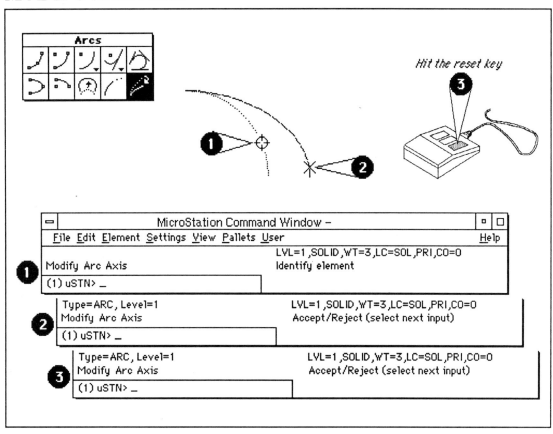

The Modify Arc Axis tool.

Polygons sub-palette tools

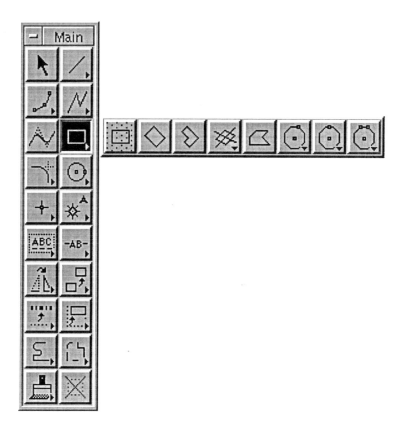

PLACE BLOCK

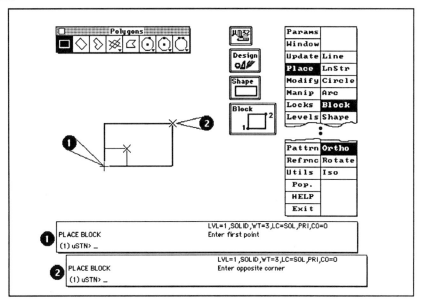

The PLACE BLOCK command...

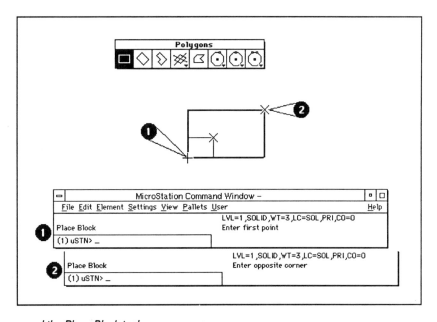

...and the Place Block tool.

PLACE BLOCK ROTATED

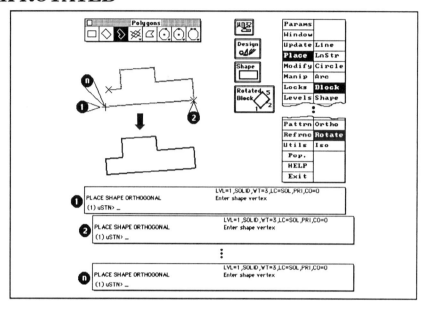

The PLACE BLOCK ROTATED command...

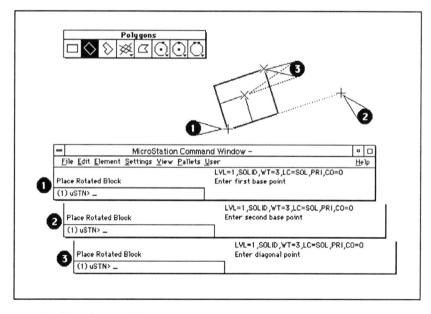

...and the Place Rotated Block tool.

PLACE SHAPE

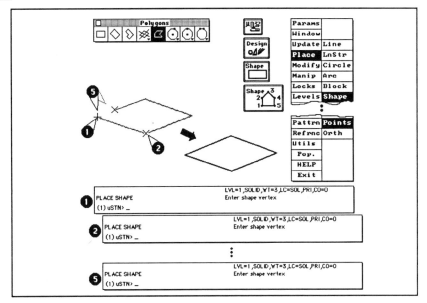

The PLACE SHAPE command...

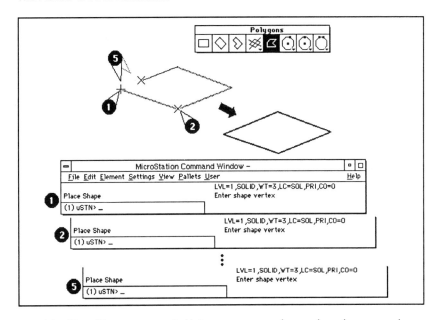

...and the Place Shape command. Note: you can now close a shape in progress by keying in CL at the uSTN prompt. This results in the shape being closed as if you datapointed back at the starting point.

PLACE SHAPE ORTHOGONAL

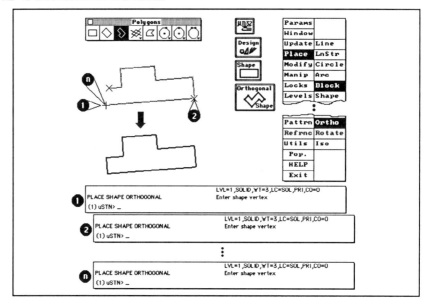

The PLACE SHAPE ORTHOGONAL command...

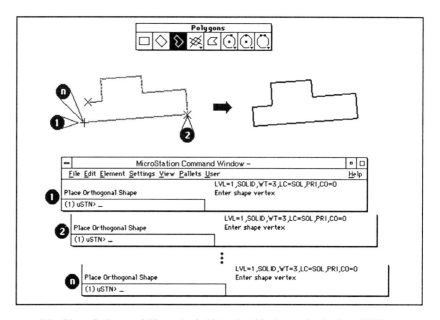

...and the Place Orthogonal Shape tool. Note: the CL closure keyin does NOT work for this tool.

PLACE ISOMETRIC BLOCK

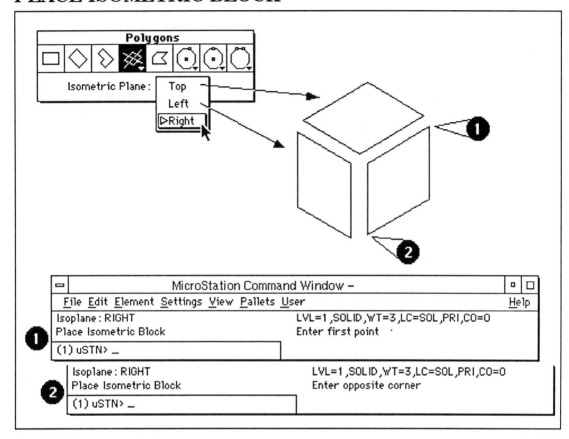

The Place Isometric Block tool. Note the pop down data field for setting which "side" of the isometric view you wish the block to be planar to. You select Top, Left or Right by clicking on the Isometric Plane field, and select the appropriate "view".

PLACE POLY INSCRIBED

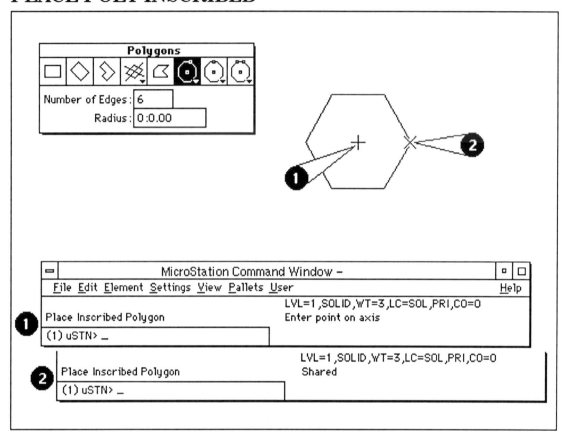

The Place Inscribed Polygon tool. Note the two pop down fields for entering the number of edges (6 in this example), and the radius at the "tips" of the polygon. When zero is entered for the radius, MicroStation prompts you for a datapoint to set this value.

PLACE POLYGON CIRCUMSCRIBED

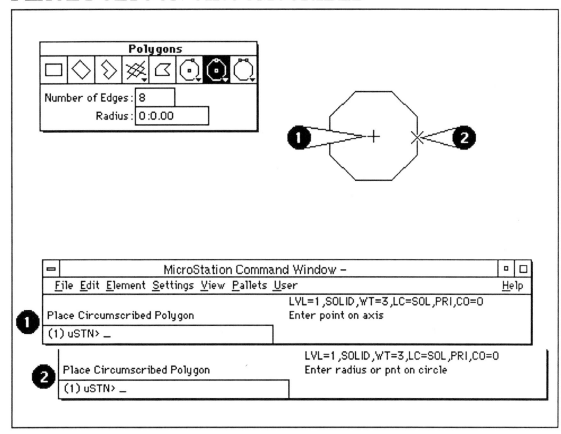

The Place Circumscribed Polygon tool. Note the two pop down fields for entering the number of edges (8 in this example), and the radius at the "flats" of the polygon. When zero is entered for the radius, MicroStation prompts you for a datapoint to set this value.

PLACE POLYGON EDGE

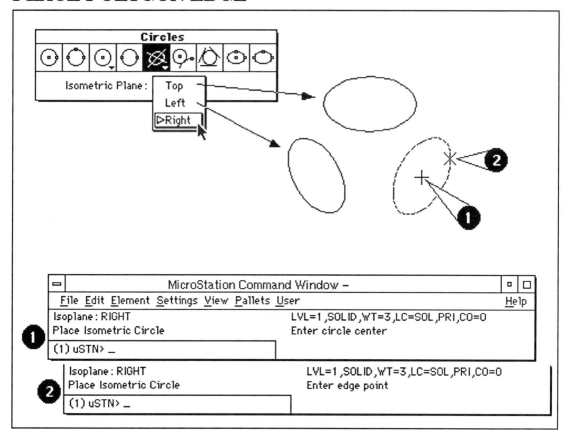

The Place Polygon by Edge tool.

PLACE LSTRING

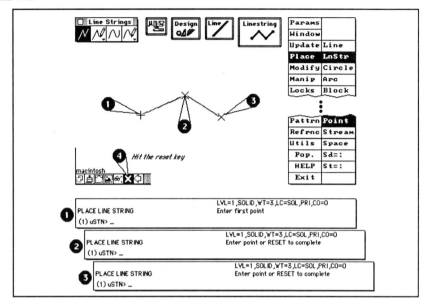

The PLACE LSTRING command...

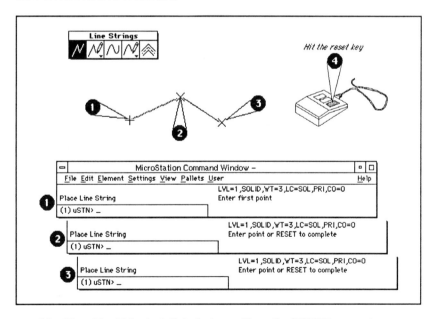

...and the Place Line String tool. Note that you still use the RESET to accept a linestring under construction.

PLACE CURVE

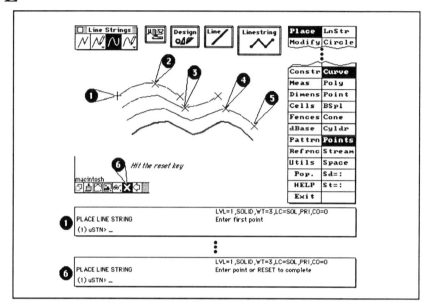

The PLACE CURVE command...

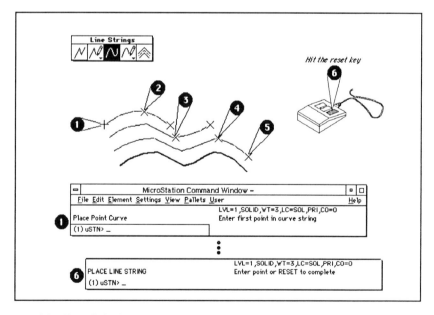

...and the Place Point Curve tool.

PLACE MLINE

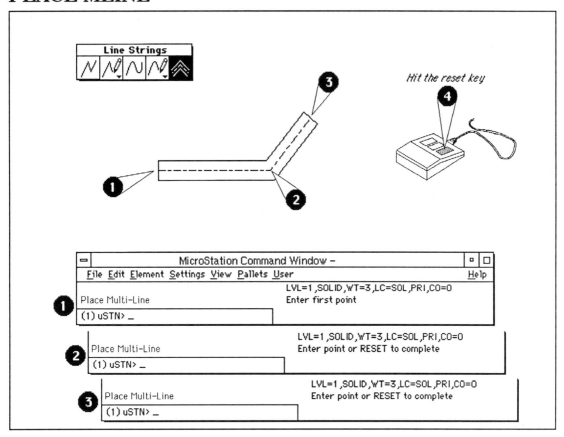

The Place Multi-Line tool. The appearance of this element is controlled by the Multi-Lines dialog box.. This dialog box is activated from the Element pulldown menu, as shown on the next page.

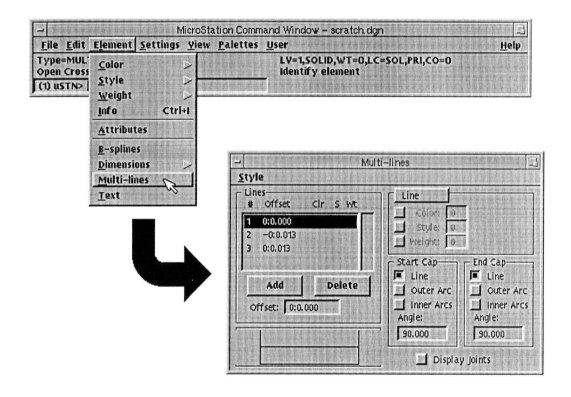

The Multi-Lines settings dialog box. From here you control the complete appearance of the multiline as it is being placed.

Lines sub-palette tools

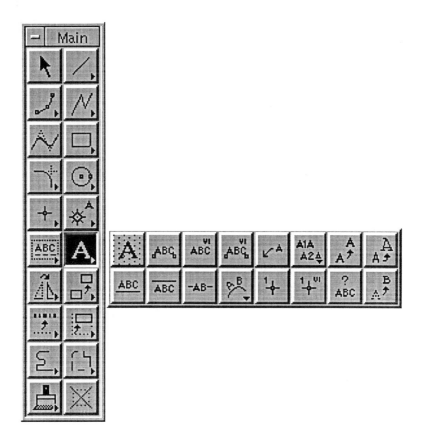

PLACE TEXT

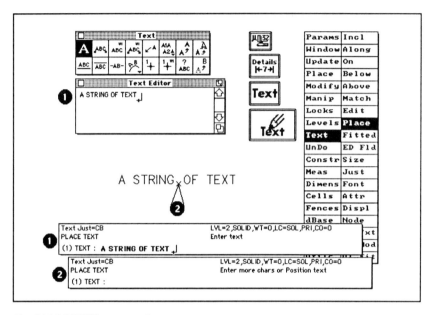

The PLACE TEXT command...

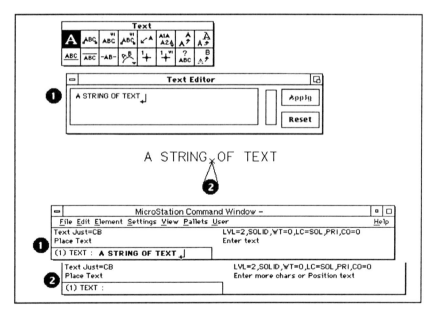

...and the Place Text tool. Note the appearance of the Text Editor window.

Text Editor window

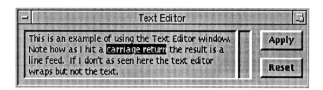

```
This is an example of using the Text Editor window.
Note how as I hit a carriage return the result is a
line feed.  If I don't as seen here the text editor wraps but not the text.
```

The text editor window.

This window is activated automatically when you select a text placement or edit text command. You enter your text for placement in this window. The insertion point for your text can be repositioned anywhere in the body of the text being inserted by datapointing at the new location. All text keyed in will be positioned between the existing text. You can also click-drag across text to delete (hitting the delete or backspace key) or replaced (typing in the new text).

NOTE: You now hit CARRIAGE RETURN ONLY to insert multiple lines of text. The text will automatically wrap around in the text editor window but will be placed according to the location of these line feeds inserted with the carriage return.

PLACE TEXT ABOVE

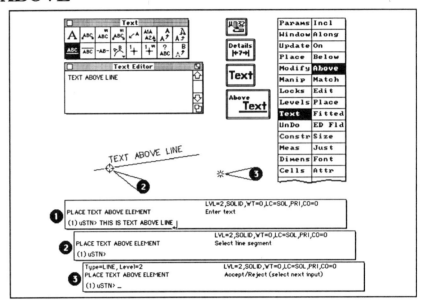

The PLACE TEXT ABOVE command...

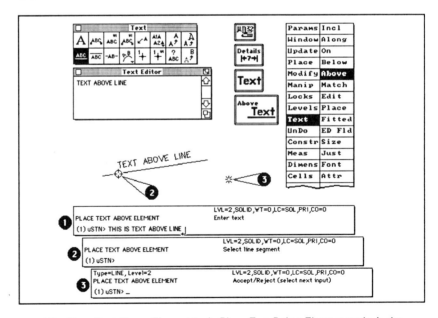

...and the Place Text Above Element tool. Place Text Below Element works in the same manner only places its text below the element selected.

PLACE TEXT ON

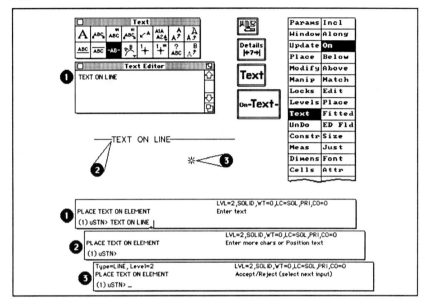

The PLACE TEXT ON command...

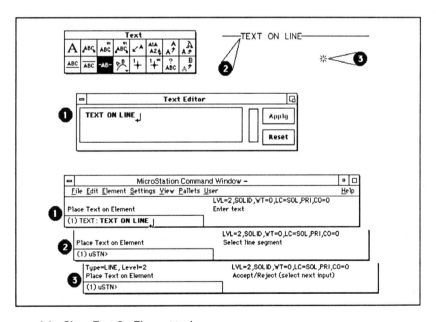

...and the Place Text On Element tool.

PLACE TEXT ALONG

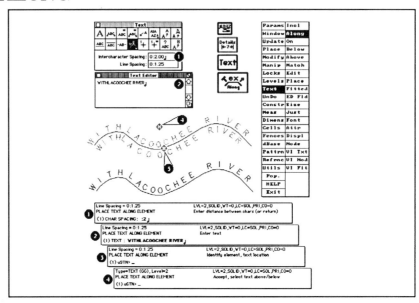

The PLACE TEXT ALONG command...

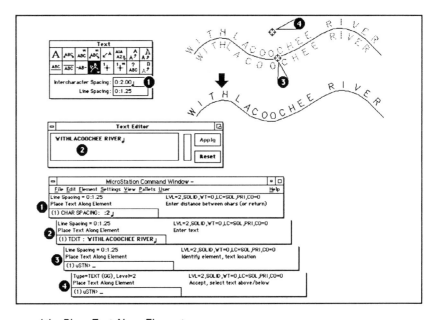

...and the Place Text Along Element.

PLACE TEXT FITTED

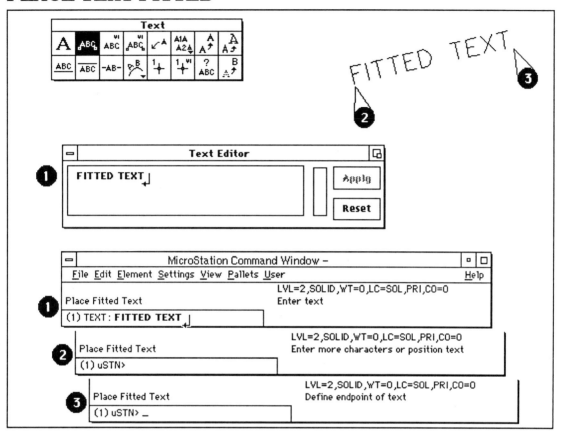

The Place Fitted Text tool.

PLACE NOTE

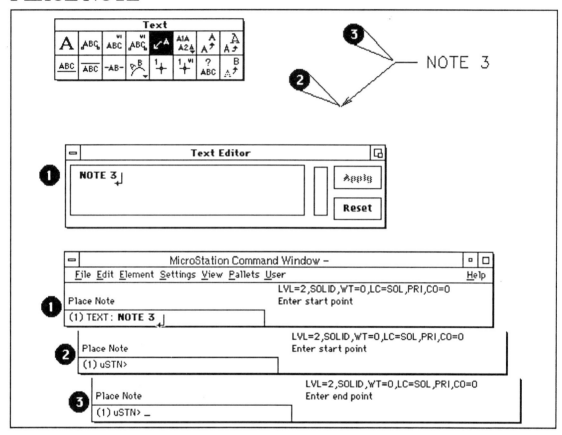

The Place Note tool. Text is optional. If none is entered, MicroStation will draw the leader line and arrowhead. This tool is controlled by the Dimension Placement dialog box accessed from the Element pulldown menu. If Location: Automatic (or Semi-Auto) is selected a single line will be drawn. If Location: Manual is selected a multi-segmented leader line is drawn. Reset accepts this line.

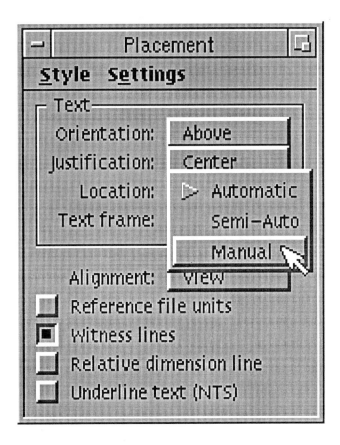

The Placement Settings dialog box for dimensioning (and note placing) control.

Copy Element sub-palette tools

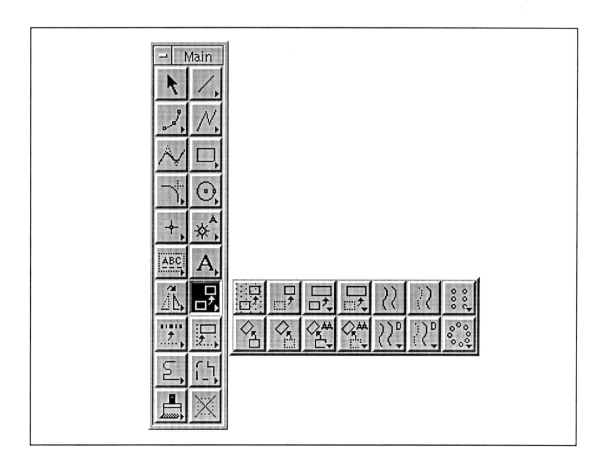

COPY

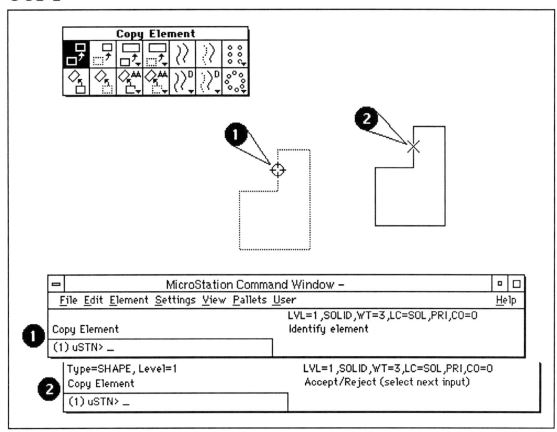

The Copy Element tool.

ROTATE (ORIGINAL)

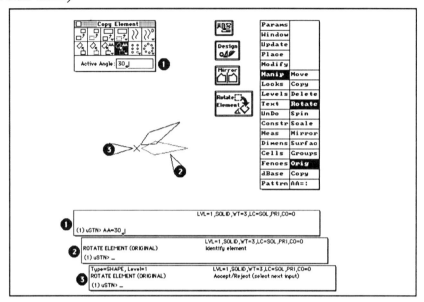

The ROTATE command...

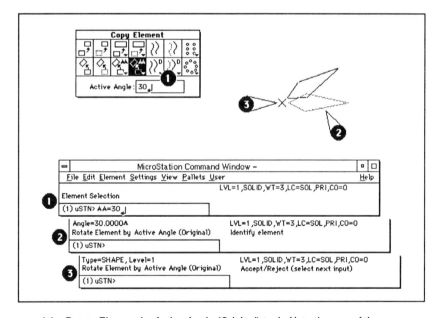

...and the Rotate Element by Active Angle (Original) tool. Note the use of the pop down data field for the active angle (AA=).

COPY PARALLEL DISTANCE

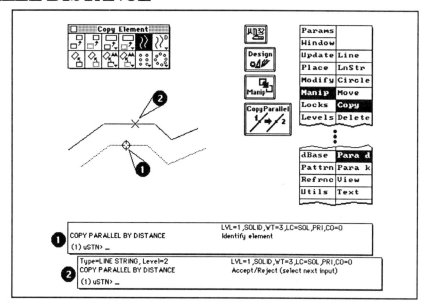

The COPY PARALLEL DISTANCE command...

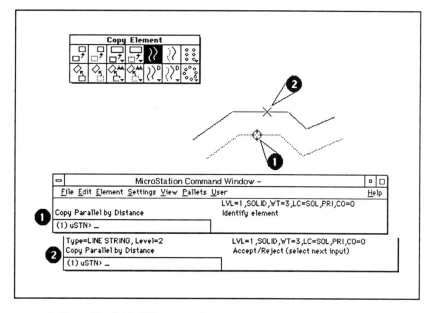

...and the Copy Parallel by Distance tool.

COPY PARALLEL KEYIN

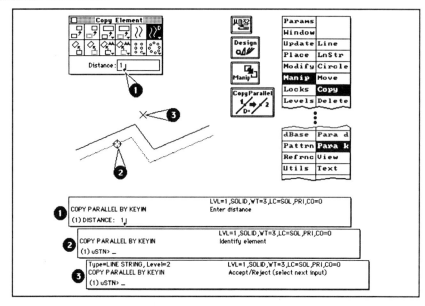

The COPY PARALLEL KEYIN command...

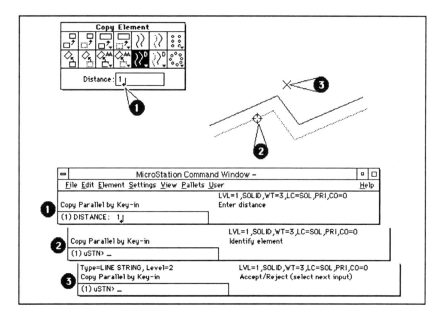

...and the Copy Parallel by Key-in tool. Note the use of the Distance popdown field for entering the delta distance.

SPIN ORIGINAL (or copy)

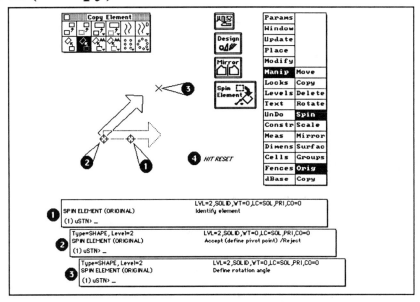

The SPIN ORIGINAL command...

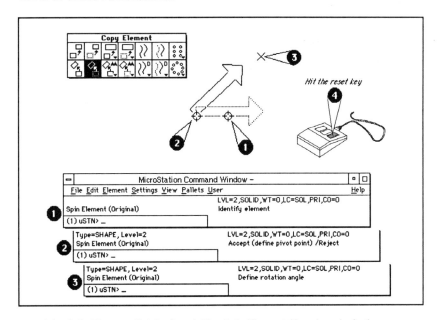

...and the Spin Element (Original) tool The Spin Element (Copy) works in the same manner as this command.

ARRAY RECTANGULAR

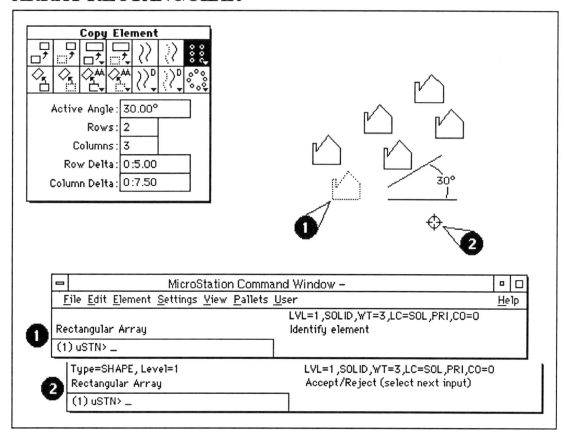

The Rectangular Array tool. Note the large number of popdown fields. These control the number of rows and columns, the angle and the spacing of these row/columns.

ARRAY POLAR

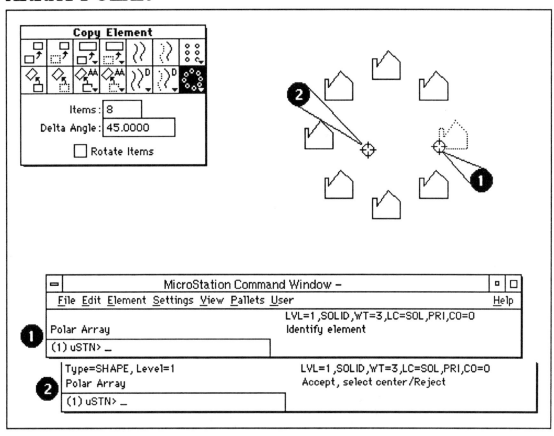

The Polar Array tool. The Rotate Items selection button controls whether the elements copied will be rotated as well.

Modify sub-palette tools

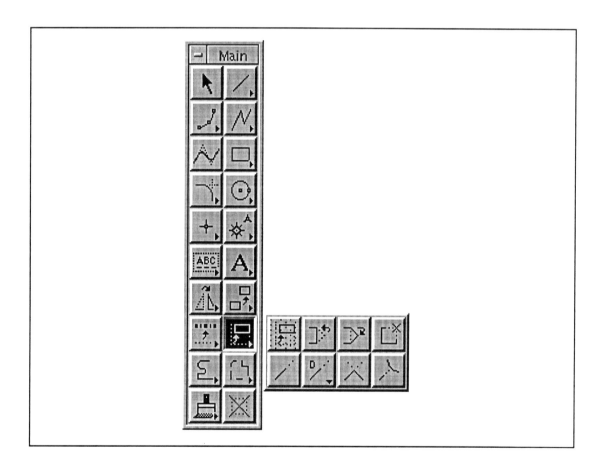

MODIFY

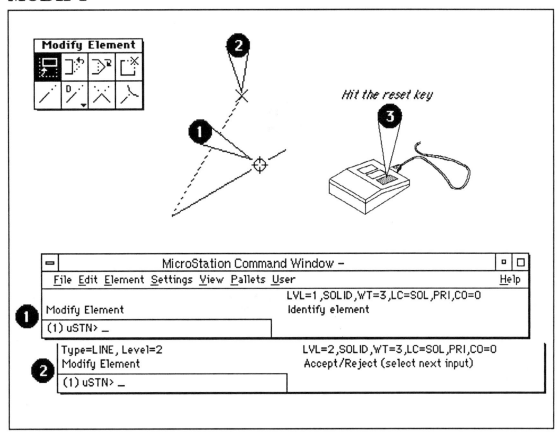

The Modify Element tool.

INSERT VERTEX

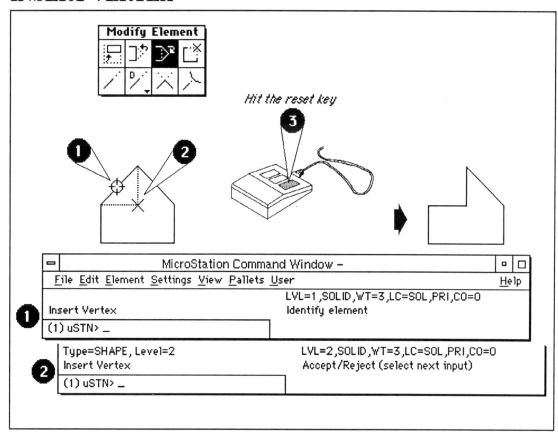

The Insert Vertex tool.

DELETE VERTEX

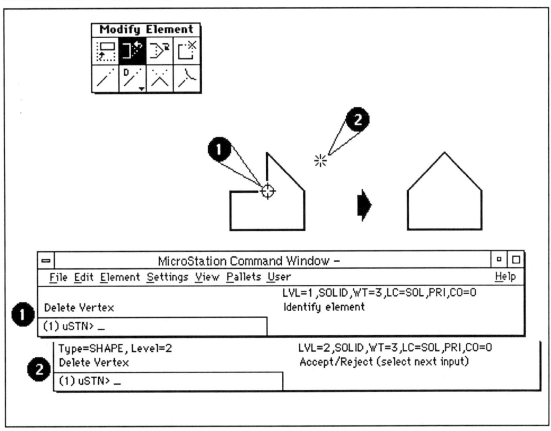

The Delete Vertex tool.

DELETE PARTIAL

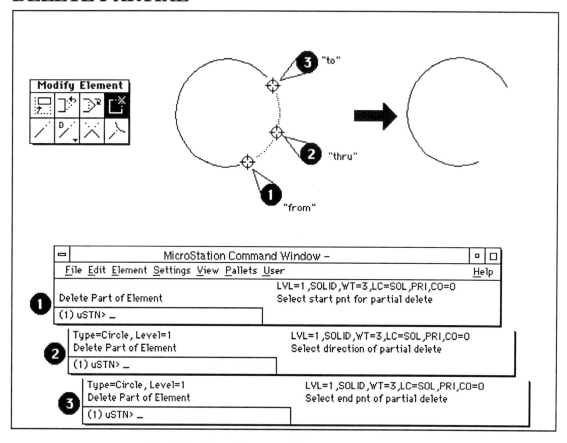

The Delete Part of Element tool.

EXTEND LINE

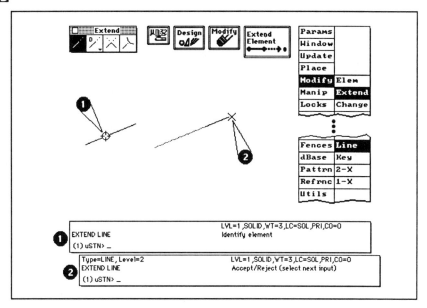

The EXTEND LINE command...

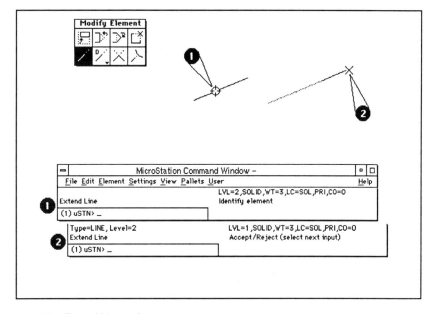

...and the Extend Line tool.

EXTEND LINE KEYIN

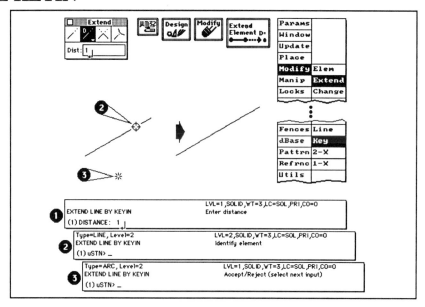

The EXTEND LINE KEYIN command...

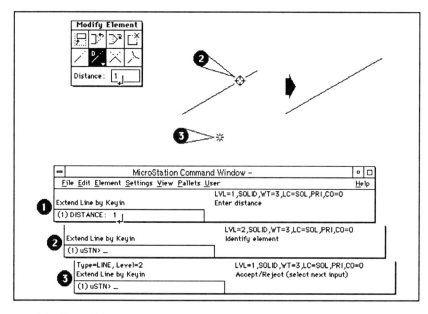

...and the Extend Line by Key-in tool. Note the popdown field for entering the distance to extend.

EXTEND LINE INTERSECTION

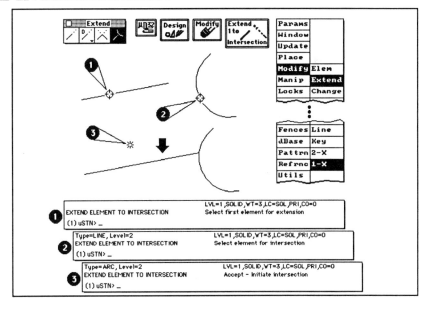

The EXTEND LINE INTERSECTION command...

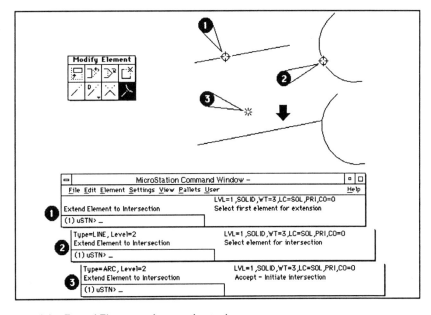

...and the Extend Element to Intersection tool.

EXTEND LINE 2

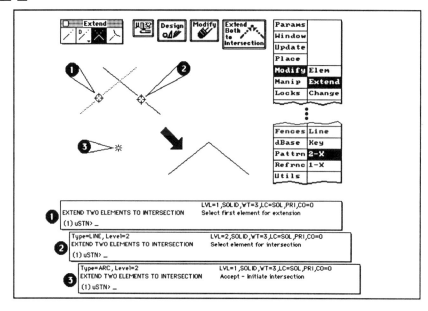

The EXTEND LINE 2 command...

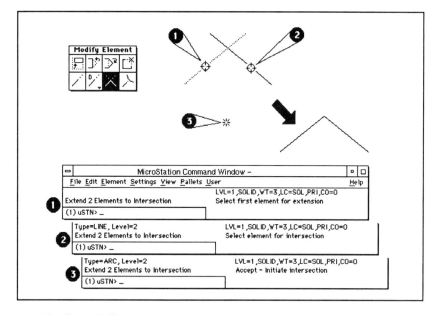

...and the Extend 2 Elements to Intersection.

Fillets sub-palette tools

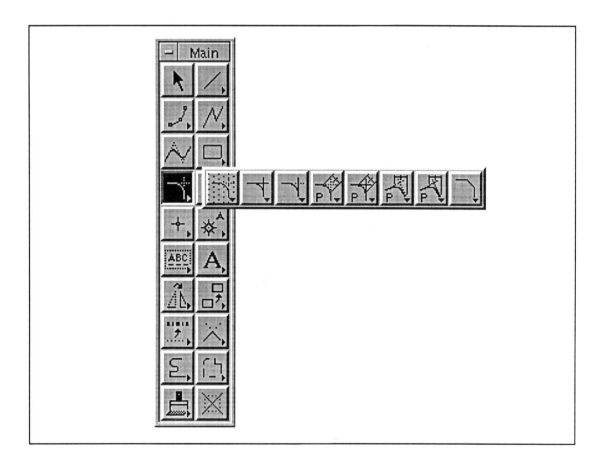

FILLET MODIFY

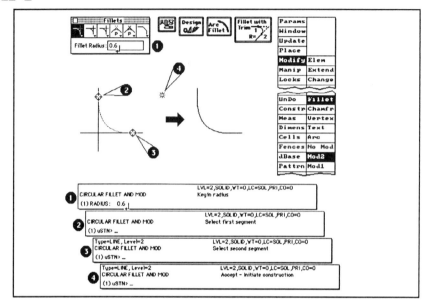

The FILLET MODIFY command...

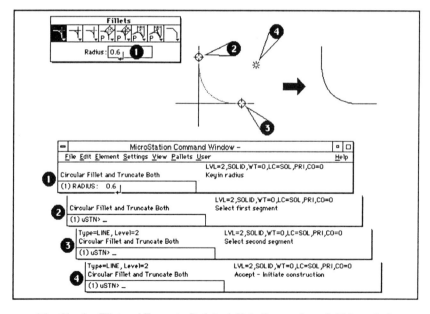

...and the Circular Fillet and Truncate Both tool. Note the popdown field for entering the fillet's radius.

FILLET NOMODIFY

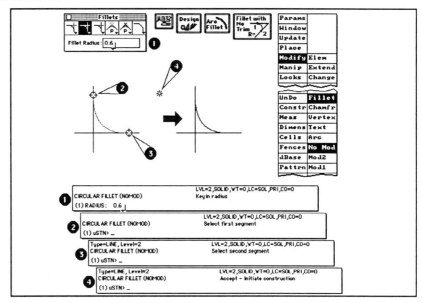

The FILLET NOMODIFY command...

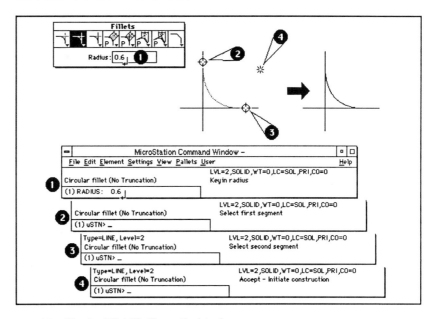

...and the Circular Fillet (No Truncation) tool.

FILLET SINGLE

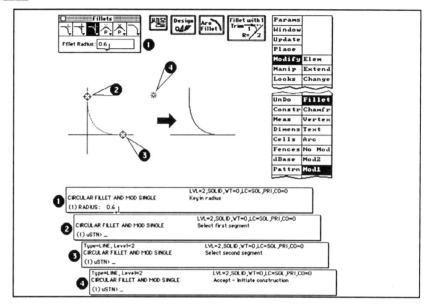

The FILLET SINGLE command...

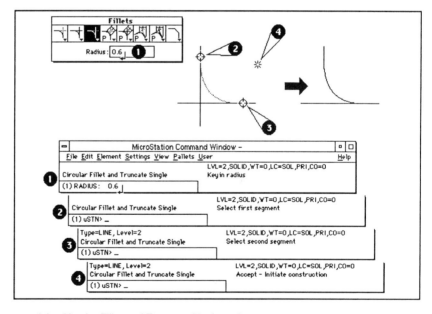

...and the Circular Fillet and Truncate Single tool.

CHAMFER

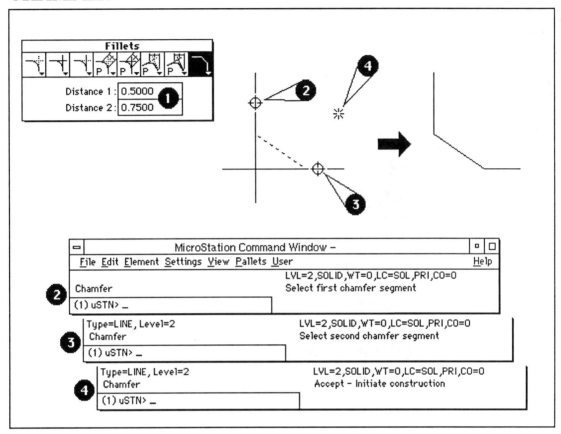

The Chamfer tool. Note the Distances popdown fields required to set the setback along the two elements selected by datapoints.

Mirror sub-palette tools

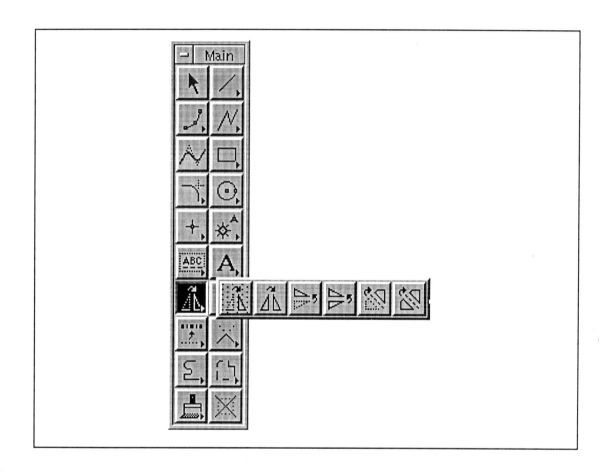

MIRROR VERT/HORIZONTAL/LINE

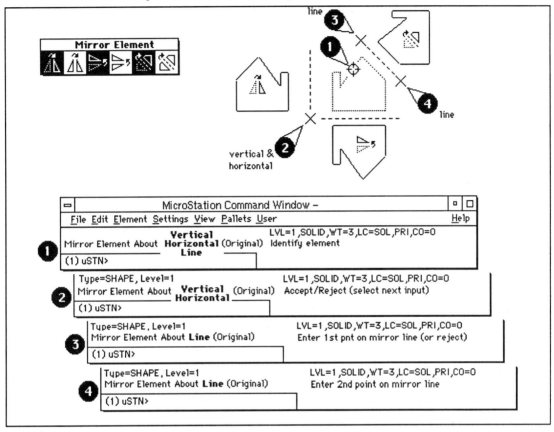

The Mirror Element About Vertical/Horizontal/Line (Original) tools. The Icon in each shape matches that of the specific tool used.

Measure palette tools

MEASURE DISTANCE ALONG

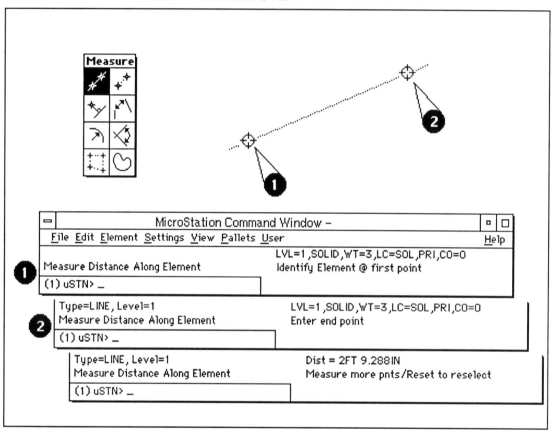

The Measure Distance Along Element tool

MEASURE DISTANCE POINTS

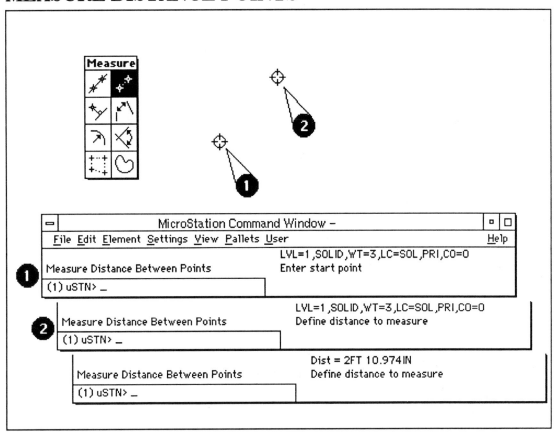

The Measure Distance Between Points tool.

MEASURE DISTANCE PERPENDICULAR

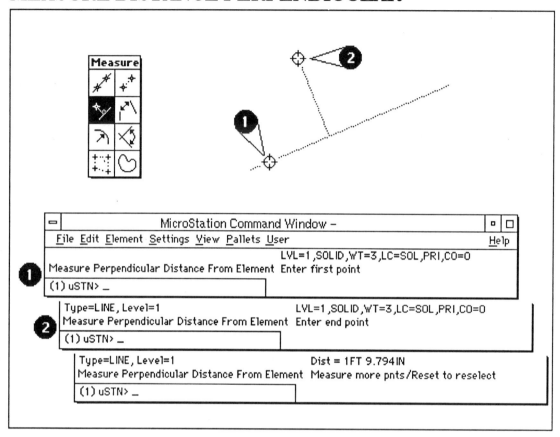

The Measure Perpendicular Distance From Element tool.

MEASURE DISTANCE MINIMUM

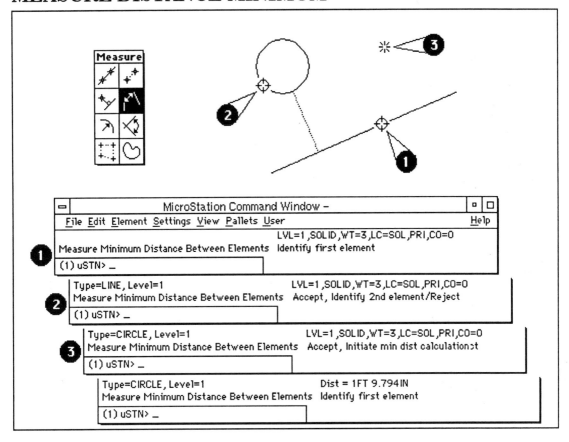

The Measure Minimum Distance Between Elements tool.

MEASURE RADIUS

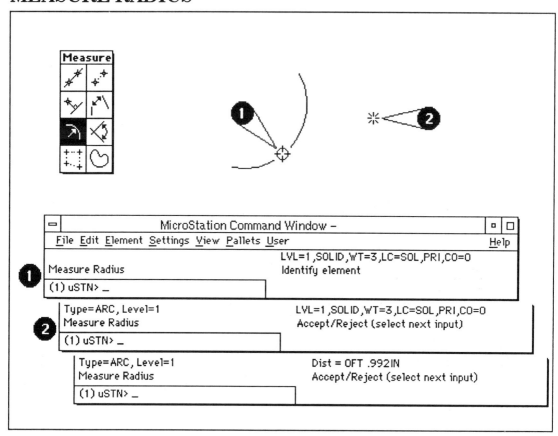

The Measure Radius tool.

MEASURE ANGLE

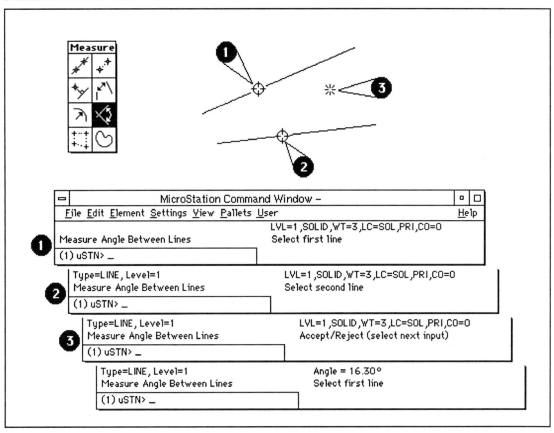

The Measure Angle Between Lines tool.

MEASURE AREA POINTS

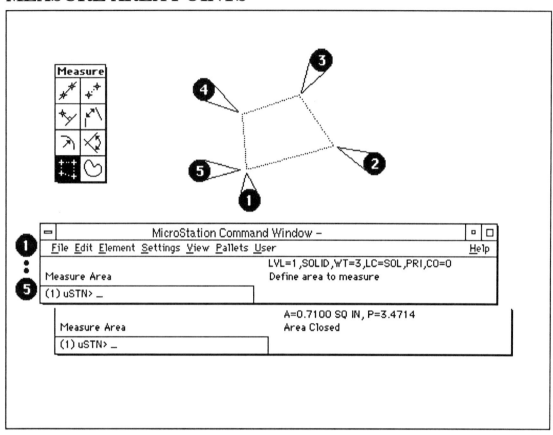

The Measure Area tool.

MEASURE AREA ELEMENT

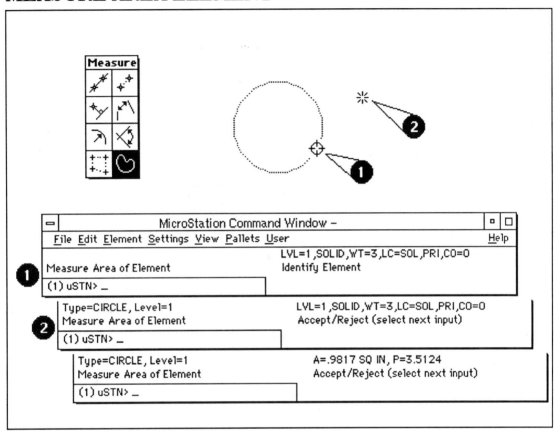

The Measure Area of Element tool.

Fence pallet tools

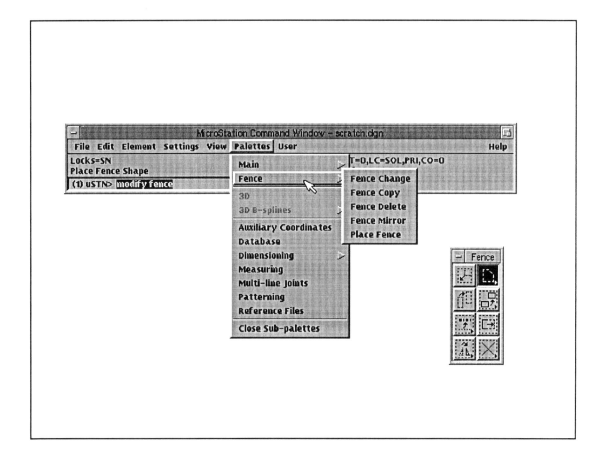

The fence manipulation tools work in the very same manner as 3.x. The various fence locks are accessed from the Settings/Locks/Full menu selection. From the Fence Selection pop up menu you select Inside, Overlap, Clip and three new choices: Void, Void Overlap and Void Clip.

Think of these last three as an inside out fence. When a fence manipulation tool is activated the Void locks cause it to operate on elements OUTSIDE of the fence area. The standard lock key-ins are still available: LOCK FENCE *INSIDE/OVERLAP/CLIP/VOID...*

PLACE FENCE BLOCK

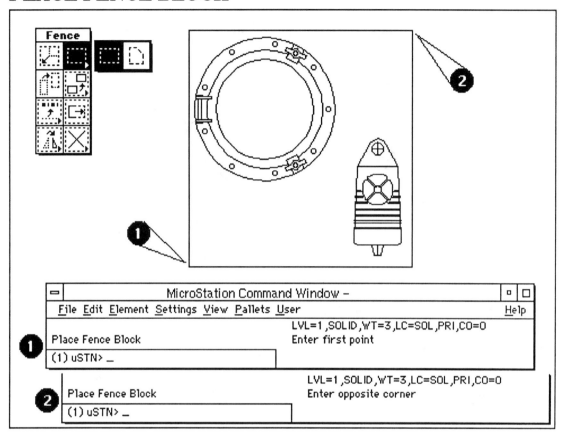

The Place Fence Block tool.

PLACE FENCE SHAPE

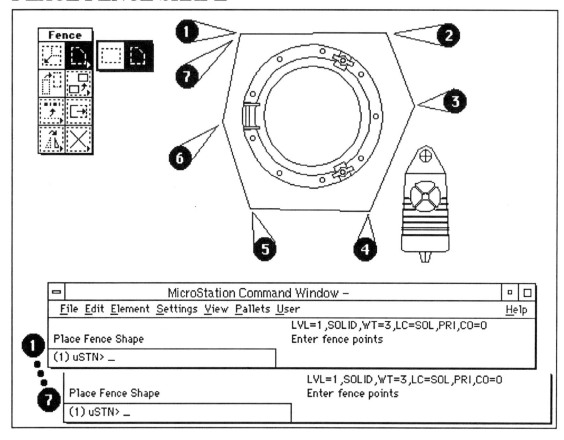

The Place Fence Shape tool.

MODIFY FENCE

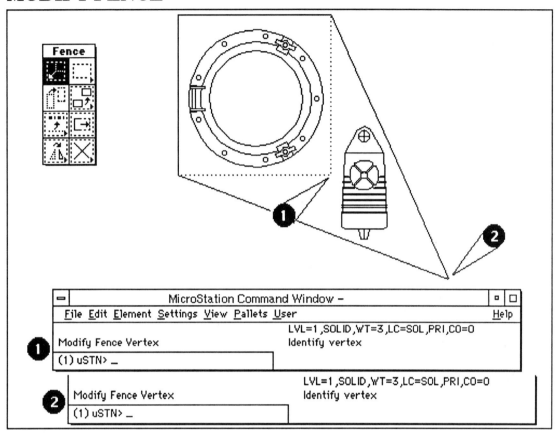

The Modify Fence Vertex tool.

FENCE COPY

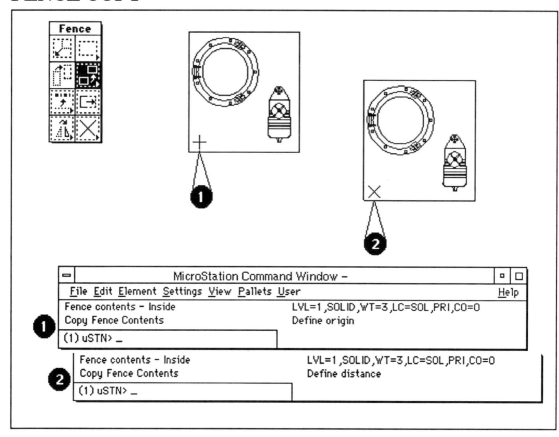

The Copy Fence Contents tool.

FENCE ROTATE COPY

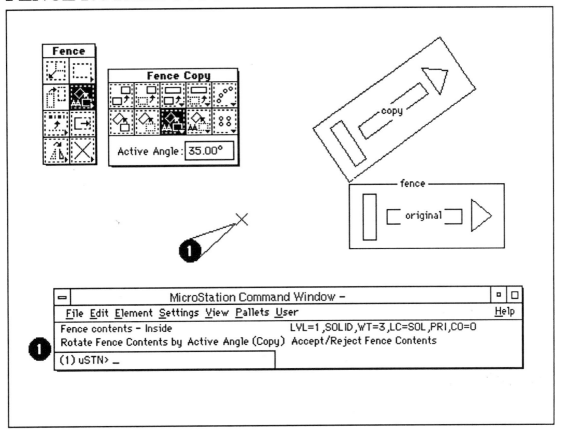

The Rotate Fence Contents by Active Angle (Copy) tool.

FENCE ARRAY POLAR

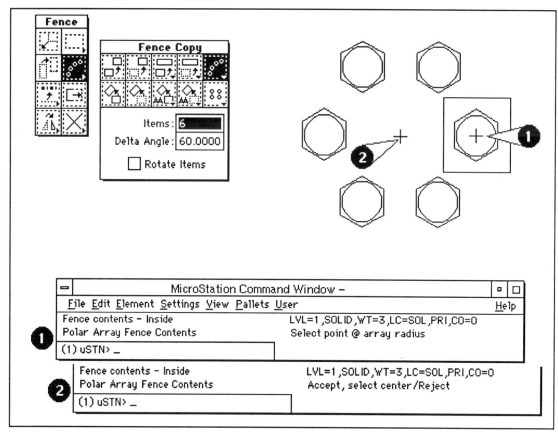

The Polar Array Fence Contents tool.

FENCE MIRROR COPY VERTICAL

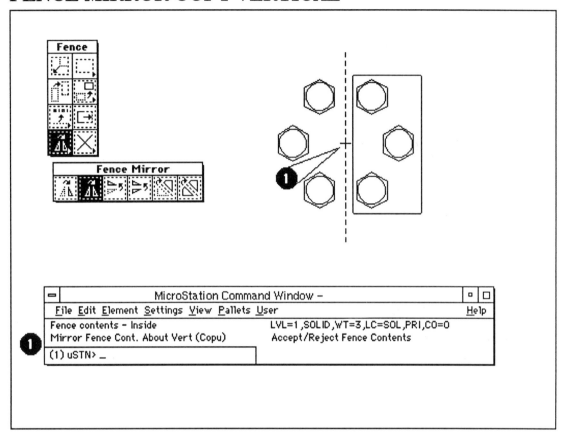

The Mirror Fence Contents About Vertical (Copy) tool.

FENCE DROP

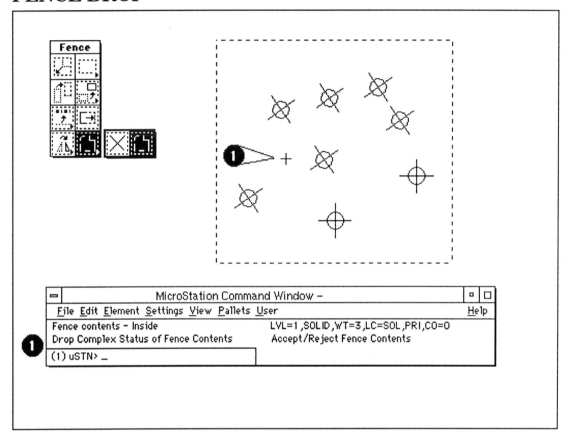

The Drop Complex Status of Fence Contents tool.

Chapter 5

Common Functions

MENUS, ATTRIBUTES AND OTHER COMMONLY USED ITEMS

This chapter shows many of the "other" functions not directly related to element placement tools. These are important and are illustrated as needed.

The Sidebar menu

The Sidebar Menu

4.0's Main Tool Palette and (surprise!) sidebar menu. Note the icon highlighted on the Main Tool palette is the same command selected on the sidebar menu.

Sidebar menus are supported by MicroStation 4.0. In fact these very capable menu types still represent the easiest method for customizing Microstation. After all, Sidebar menus are nothing more than text files.

Design Options

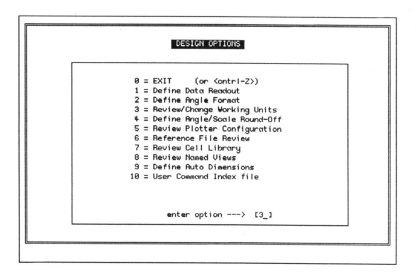

Design options under 3.x...

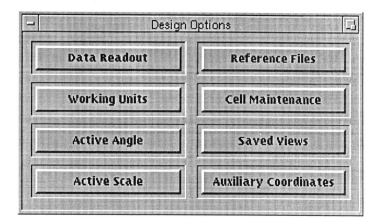

...Keying in Design Options gives you this simple dialog box. Clicking on any of the options invokes the settings box associated with the pulldown menu of the same name.

Working Units

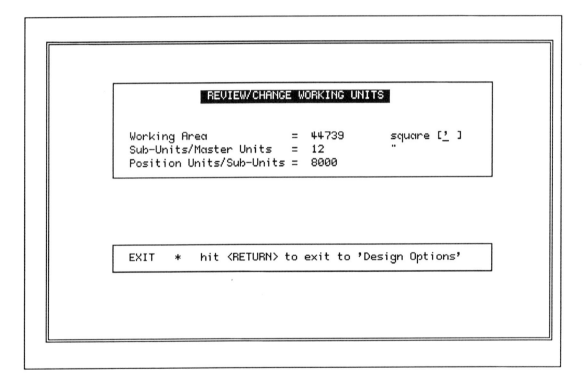

The original method for setting your working units...

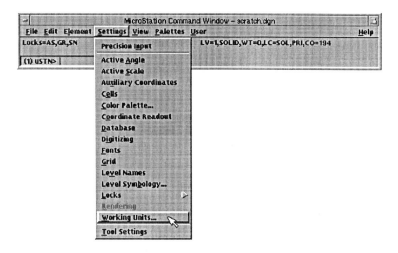

Selecting Working Units from the Settings menu gives you...

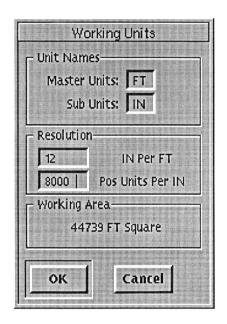

...this dialog box. You select your unit labels first, then your MU:SU:PU relationships.

Cells

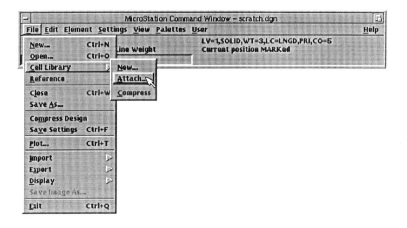

Attaching a Cell library via the File menu.

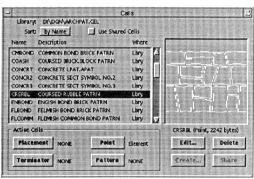

The very capable cell library settings window. From here you can perform most of the cell library maintenance duties. When a cell is selected, a "thumbnail" view of it is shown in the view box on the right.

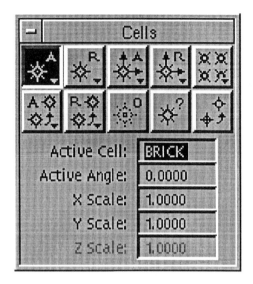

The Place Cell Absolute tool...

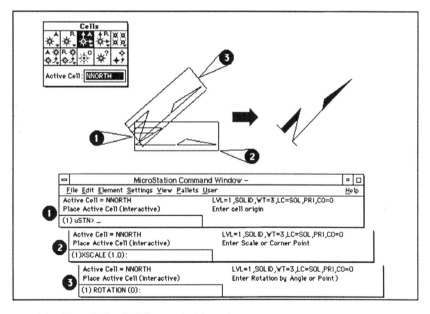

...and the Place Active Cell (Interactive) in action.

Text Attributes

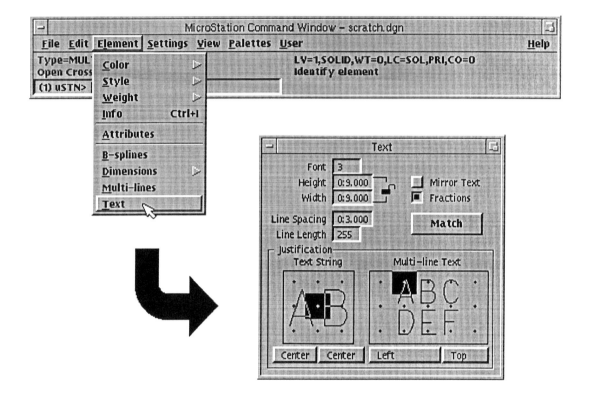

Text appearance or Attributes have been combined in one settings window.

Patterning

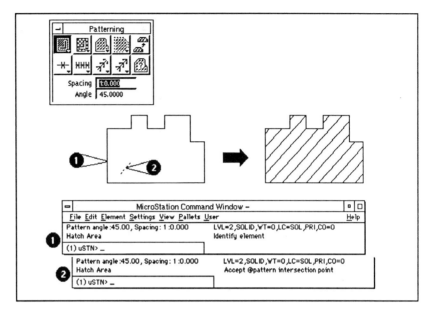

The Hatch Area tool...

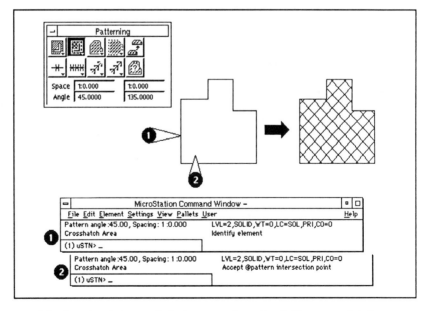

...and the Crosshatch Area tool. Both use popdown data fields to control their operations.

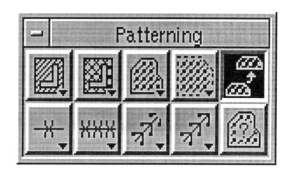

The Match Pattern Attributes, a very powerful query command. By identifying a previously patterned area you can set up your next patterning command to match the results.

Element Symbology Control

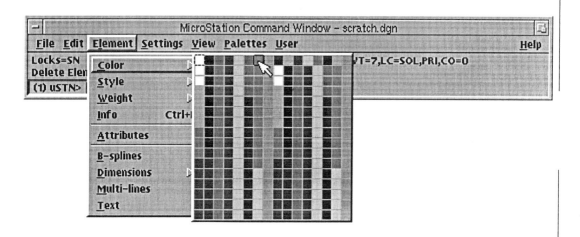

Selecting active color could not be easier.

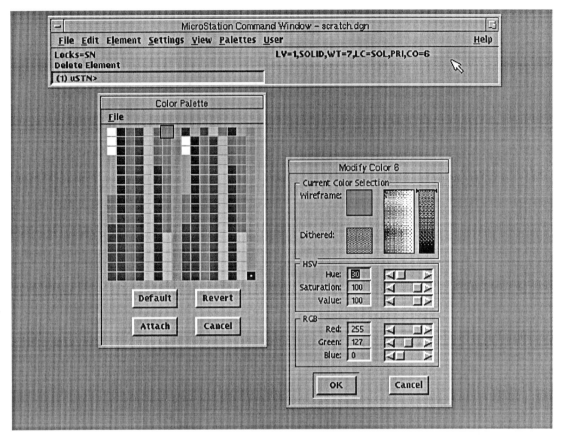

...or simpler to modify. The sliders allow you to adjust each color gun (RGB) or pick a color from the spectrum bar.

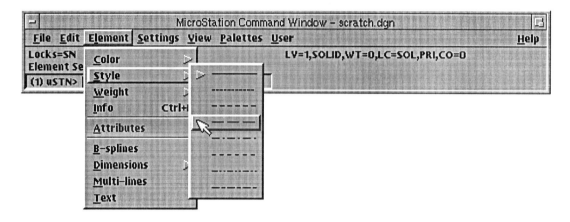

Select Active Linestyle pulldown menu selection.

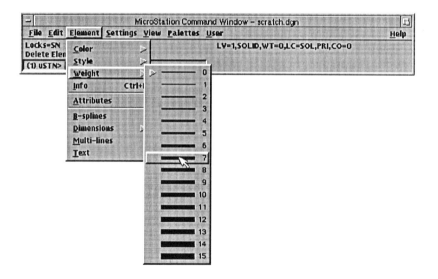

Select Active Weight pulldown menu selection.

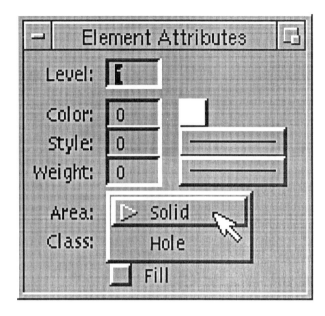

The Element Attribute dialog box. A one stop box where you can set all of the element attributes.

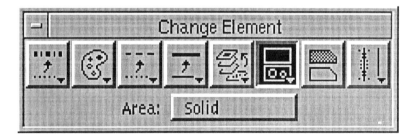

The Change Element sub-palette. All of the change symbology commands reside here as well as Solid/Hole, Class and Fill control tools.

Plotting

Plotting is controlled by its very own dialog box. Selected from the File pulldown menu you are presented with the following:

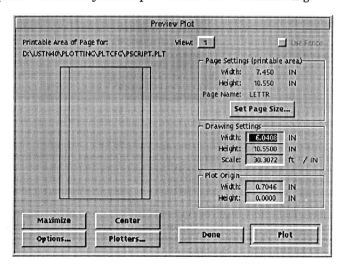

The Plotting dialog box. You set your paper size, and other parameters via this menu. In addition, you can reselect your active plotter via the Plotters... button.

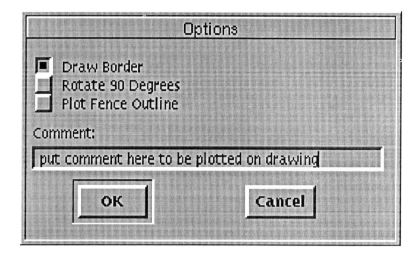

The plotter options pop up dialog box.

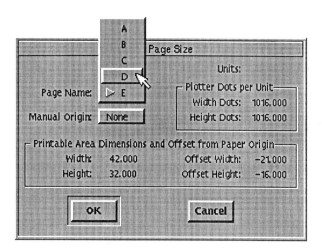

The Set Page Size button on the main Preview Plot dialog box brings up this window. A 'D' size drawing is being selected from the Page Name popup menu.

Chapter 6

Utilities

MCE, MICROSTATION MANAGER AND OTHER TOOLS.

With all of this emphasis on graphics, one has got to wonder what happened to all of the functions performed by MicroStation Command Environment? The answer is they have been incorporated into MicroStation itself.

This does not mean MCE is gone. It is still being delivered with the 4.0. However, this may not be the case with future releases. Intergraph is moving away from MCE. To replace it, MicroStation Manager has been developed.

An MDL application, MicroStation Manager is responsible for a number of functions including file creation, plot file control, and file maintenance (copying, deleting, merging files).

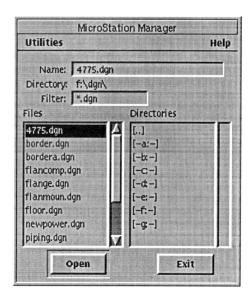

MicroStation Manager, the "New" MCE.

MicroStation automatically comes up when you key in USTATION without a file name... assuming it has been assigned as a start up application (the default).

When you run USCONFIG one of the choices is the User Preferences section. Under this section you are presented with the opportunity to specify a startup MDL application. The default is mm.ma, the MicroStation Manager. If this is not the case then MicroStation will be unavailable to you.

The MicroStation Manager is selected as the startup MDL application via this screen.

The "Other" Utilities

Importing and Exporting files

The "other" function performed by MCE is that of translator. There are a number of file translation utilities bundled in MCE that are necessary for the day to day production of drawings. Fortunately these translation duties have been picked up by MicroStation itself.

To support the bidirectional nature of file translation, the Import and Export menu selections (from the File pulldown menu) provide you with access to the DXF option. Selecting this option presents you with a dialog box for selecting the import file, which should already exist, or for keying in the name of the export file, which should not exist...yet.

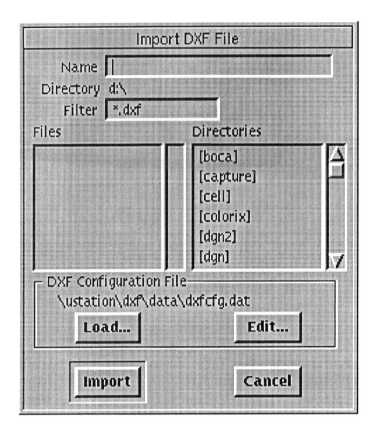

The DXF Import dialog box.

On the Import/Export dialog box is a section dealing with the DXF Configuration File. The default table that controls all of the various parameters is DXFCFG.DAT located in the \USTATION\DXF\DATA directory (the Intergraph people really love subdirectories).

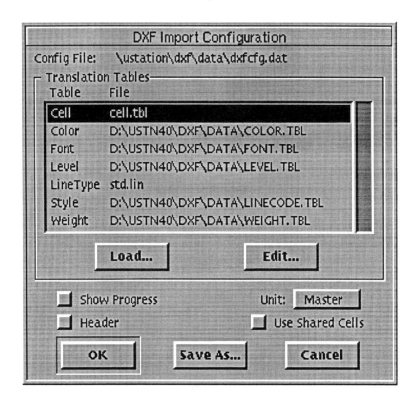

The configuration file that controls the DXF file process.

By selecting the Edit... button you are presented with the DXFCFG.DAT parameters. Each table shown is yet another file on the system located in the same directory as DXFCFG.DAT. By clicking on and selecting the Edit... button again, you gain access to this final level of DXF control. For instance selecting the Cell table presents you with the Edit DXF Translation Table window.

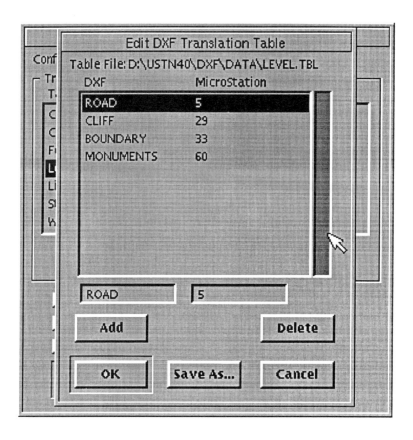

The Edit DXF Translation Table dialog box. You can edit the individual parameters by selecting them with the mouse and typing in the data. As with 3.3 these control files are simple text files, editable with any text editor.

ns of data   6-177

Exporting other types of data

2D to 3D and back

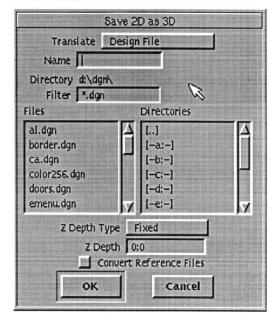

You select the name of the output file and how the Z axis is treated in the corresponding file.

The Renderman connection

One of the most fascinating options of MicroStation is its ability to interface with the Pixar Renderman rendering software.

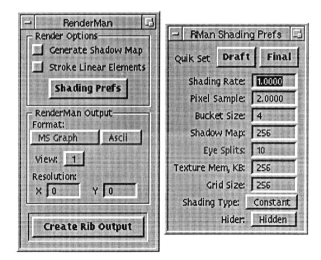

The RIB file

The result of selecting the Rib option from the export command will be the Renderman dialog box. From this dialog box you set the numerous options from the type of output required (MicroStation graphics to color Postscript) to the type of shadows you wish cast.

Visible Edges

An output capability closer to home is the export visible edges. Meant for a 3D file, this option creates a "flat" file of a view with the hidden lines either removed or dashed.

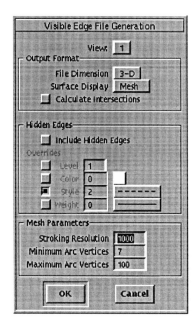

The dialog box used to control the outcome of a visible edges creation.

A Word about the future

Intergraph has stated in their certification notes that MCE, tutorials, USTN sidebar menu and even the venerable command menu (MENU) will not be included in future releases of MicroStation. It is their intent to phase out these "links to the past" and encourage the use of the new tools introduced with 4.0.

Although this sounds rough at first, the truth of the matter is that the future is with these other tools. Although Intergraph may continue to maintain the ability to use command menus and sidebar menus, it will be up to the individual MicroStation users to update these files for the future.

Chapter 7

Updating your system to 4.0

SOME CONSIDERATIONS

MicroStation 4.0 is a very powerful package and may be a bit much to take in one swallow. A few helpful suggestions follow to ease the transition a bit.

Running concurrent versions (PC version only)

You may want to run both 3.x and 4.0 on the same machine during your transition from one to the other. To accomplish this you can install MicroStation 4.0 in a directory other than \USTATION. By using two separate PATH statements (stored in short batch files, of course) you can switch between the two packages with ease. Some caveats about this are in order.

If you intend to use 4.0's unique elements in drawings destined for 3.x, you should either SET COMPATIBILITY ON or FREEZE elements.

The space required to house 4.0 is considerable. Approximately 22MB are required to hold all of the various options (demo files, patterning utilities, etc.).

Before you begin the process you should evaluate saving any of the files you have customized:

☐ \USTATION\DATA\ - any user modified seed files

☐ \USTATION\DATA\UCONFIG.DAT

☐ modified MCE files

Developing Dimensioning Control

Probably one of the first areas that will get out of control is the dimensioning system. Due to the large number of options present in the dimensioning operations it is very easy to end up with drawings having inconsistent dimension appearances. Use the STYLE capability to contain this problem and provide a company standard dimensioning style.

Hardware Considerations

4.0 is the first version of MicroStation to run exclusively on a 80386 (or 486) based computer. This was necessary to provide the maximum performance to the user and take full advantage of all memory supplied to MicroStation. MicroStation uses a fair amount of memory; 2MB is considered minimum to run the main executable. However to really take advantage of its capability, 4 to 8 megs is more appropriate. As with 3.3, a virtual disk increases performance.

If you run an expanded memory manager software program on your system, you will want to make sure it is VCPI compliant. This fact is necessary to assure that Intergraph's memory manager (PharLap) can coordinate its use of the system's memory with that of this memory manager.

If you plan to run MicroStation under Windows 3.0, you should read the release notes about configuring the various software packages. Found in \USTATION\DOCs, these files are very good at explaining how to do this...at least as good as can be expected.

Video Considerations

MicroStation 4.0 is, by far, the most graphically oriented CAD package available. For this reason some thought should be given to the video capability of your system. VGA is the most common video available for the PC. A number of VGA manufacturers have begun to support "super VGA" cards capable of displaying greater numbers of colors and more dot resolution. A standard has been developed that allows these super VGA cards to interface with advanced products like MicroStation. Called VESA, this standard is either built in to the video card or is loaded in at time of boot.

MicroStation supports the VESA standard and will allow you to configure the video to take advantage of it. Additionally, MicroStation supports advanced video cards such as IBM's 8514 and the TIGA (Texas Instruments Graphics Architecture) equipped cards. The most apparent benefactor of these advanced cards is the dynamic panning feature. Panning becomes much faster.

These are just some of the factors to consider when upgrading to 4.0. Just remember, no matter how intimidating and different this new software appears, underneath it all the solid good 'ole MicroStation routines (and IGDS before that) are still there. This is one time the adage "all you know is wrong" does NOT apply!

Chapter 8

MicroStation For The IGDS User

A STUBBORN IGDS USER'S GUIDE TO MICROSTATION

So you sit there in front of your InterAct workstation banging away at your current job. Recently, while reading Intergraph's "propaganda" (a.k.a. sales literature) you became VERY interested in this new MicroStation 4.0. Not looking a thing like the IGDS you've become accustomed to you, wonder if its for you. Well, the answer is ... YES!

IGDS, the venerable "grandfather" of Intergraph CAD, is still a VERY capable CAD system. MicroStation, in fact, did not support all of the features of IGDS until version 4.0. This being the case you, the dedicated IGDS user is probably in a better position to utilize all of the features of 4.0. You can appreciate the addition of dynamic panning and polygon reference file clipping.

Similarities Between MicroStation and IGDS

So, where to begin? MicroStation 4.0 is, first and foremost, an Intergraph product and as such does maintain a high level of compatibility with IGDS. In fact all design files created on an Intergraph Vax IGDS system are immediately usable. The same goes for cell libraries, most user commands and menus (with some modification). This means you do not sacrifice your extensive and expensive design file archives you've developed over the years.

So, this leaves you with a system that takes the best of IGDS and adds the latest user interface technology. In addition a number of new elements have been added to complement the existing IGDS "core".

The Vax-based Intergraph Command Environment (ICE) is used for performing most non-graphical functions. Here you create design files, cell files, compress, reduce, merge, and otherwise manipulate the design file. MicroStation performs most of these functions from the MicroStation Manager. When you invoke MicroStation (via the USTATION keyin) without a filename MicroStation Manager is automatically activated.

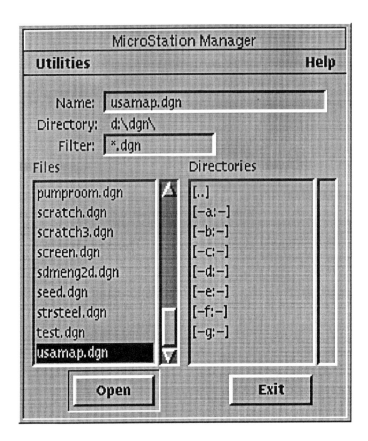

MicroStation Manager, the graphic equivalent to the Intergraph Command Environment.

Where MicroStation and IGDS Differ

Previous versions of MicroStation more closely echoed the "look and feel" of IGDS. The command window displayed the four message fields the same information as IGDS: PRompt, COmmand, ERror/Keyin, and INformation. The display even split into four views just as on the InterAct workstation.

MicroStation 4.0 retains the command window. However the use of the ERror/Keyin field has changed dramatically. IGDS and earlier versions of MicroStation used this field to prompt you for additional parameters such as an arc's radius or a copy parallel distance. Under 4.0 this information is entered as part of the tool palette in the form of popdown data fields.

Probably the biggest change in store for IGDS users is the use of these tool palettes. This is a totally new method for selecting a graphics command. True, MicroStation 4.0 still supports command menus. But the real power to this new product is the palette tool system. Truthfully you may be at an advantage over a MicroStation user in that the key-in commands were never available to you.

Think of these tool palettes as a cross between an on-screen command menu fragment/tutorial. Probably the biggest hurdle with palettes is finding where a particular tool is located (remember memorizing the command menu?). The good news is the organization of the tool palettes is much more logical than the command menu. Line construction tools are grouped together, circles with circles, arcs with arcs, etc.

You can, of course use the command menu (AM=MENU,CM) with 4.0. However, all of the newer commands will not be found there, and future versions of MicroStation will NOT include the command menu!

The Databuttons

Although you have the same four button assignment on your digitizer puck the function of one of them has a new wrinkle. The tentative button can now identify element intersections and multiple keypoints along an element. Intersection snap is controlled either by a key-in (LOCK SNAP INTER) or a check button on the Full Locks dialog box (from the Settings pulldown menu). When active, this snap will put the tentative point precisely at the intersection of two elements.

Multiple tentative points are used to identify a series of equal spaced points along an open element. Using the KY= key-in or the Tentative points data field on the same Full Locks dialog box, you set how many segments the element should be "broken" down to. KY=4 results in four quarters, thus a total of five keypoints (the two ends and each quarter).

View Manipulation Commands

Gone is the "sweet spot" found in the middle of every InterAct workstation. Instead you control the functions of each view by identifying it directly (datapoint in it) or through the use of a pop-up menu. For instance, to turn a view on you select the Open/Close command from the View pulldown menu. This leads to a popout menu from which you identify the view you wish to use.

The View Attributes dialog box would be equivalent to the view control commands on the command menu. Here you select the appropriate view in the View Number selection field. As you toggle the various display features such as grid display, fast font, etc, you Apply them to the view in question by hitting the Apply button. To simulate the sweet spot, the area of the screen that affects all views, there is the All button. Hitting this does as its name implies, sets all of the views to the toggled values shown.

By now you have seen how views can be resized and repositioned all over the screen. There is a command available that "tiles" the views as you would expect to see them in IGDS. The Tile command is available to do just this (also from the View pulldown menu). The order of the views by number are a little different: the default is clockwise from the upper left display corner: 1,2,3,4. You can set a user preference check box to simulate IGDS (counterclockwise from the upper left).

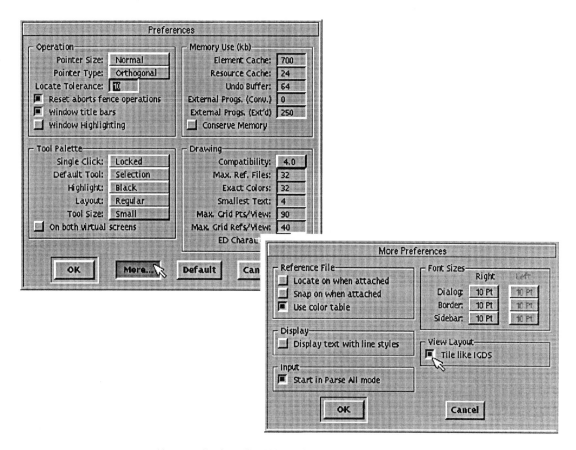

You can tile views like IGDS using the More Preferences dialog box.

Panning around

MicroStation 4.0 introduces a new/old feature: dynamic panning. By holding the shift key down while datapointing and sliding you can pan around in a design file in a manner very similar to the dynamic panning function of the InterAct workstation. The further you move the cursor from the initial datapoint the faster the panning action appears.

In the case of MicroStation, however, there is no limit to the dataset size. In other words you don't have to download anything to make this feature work! It does, however, require certain video cards to operate. See the MicroStation Installation guide for more information.

Design Options Tutorial, or the Lack Thereof

You can forget about the old reliable Design Options Tutorial. Instead, you select the various functions from the pulldown menu bar. Many of them are found on the Settings pulldown menu, others on the File pulldown menu and yet others on the Element pulldown menu.

Design Options tutorial versus the appropriate dialog boxes

Design Option	Pulldown menu/command
CELL LIBRARY	Settings/Cells
REVIEW/CHANGE LEVEL SYMBOLOGY	Settings/Level Symbology...
DEFINE DATA READOUT	Settings/Coordinate Readout
DEFINE ANGLE FORMAT	Settings/Coordinate Readout
AUTOMATIC DIMENSIONING	Element/Dimensions
DEFINE ANGLE/SCALE ROUNDOFF	Settings/Coordinate Readout
REVIEW/CHANGE WORKING UNITS	Settings/Working Units...
DEFINE DMRS DATABASE	Settings/Database*

The Plotting Menu

The Intergraph Plotting System is GONE!. Instead, you are presented with the much better organized Preview Plot dialog box. The functions are still the same, with working units versus plotter units. Instead of selecting a destination plotter you identify the type of plotter you wish to process a spool file for.

If, for instance, you have a number of different plotters to support, you merely select the appropriate plotter type from the Select Plotters dialog box (from the button of the same name). Selecting the Plot button prompts you for a file name to which the spool data is written.

To actually send the data to the plotter you use either the MicroStation Manager's Plot file function or an operating system copy or print command.

For more detailed information on plotting, see *Inside MicroStation*.

In A Word...

MicroStation 4.0 is first and foremost an enhanced IGDS product. All of those good design habits you've developed over the years still apply. In fact, with respect to some "new" tools you already know them well. The "nice" graphics display, although a very power tool in itself, does not take away from fact. When it gets right down to it MicroStation still requires you to be the brains behind the datapoint. So relax! MicroStation 4.0 is just another fantastic tool to make your job go faster.

Index

!

2D to 3D and back 177

A

Arcs sub-palette tools 85
Associated Dimensions 36
Associative dimensions 6

B

B-sline commands 48
B-spline commands 11

C

Cells .... 160 - 161
CHANGE MLINE 43
Circle/Ellipse sub-palette 74
Command Window 15, 17, 19, 21, 23
Constant shading 53
CONSTRUCT BISECTOR 60
CONSTRUCT BISECTOR LINE 61
CONSTRUCT LINE AA1 65
CONSTRUCT LINE AA2 67
CONSTRUCT LINE AA3 66
CONSTRUCT LINE AA4 68
CONSTRUCT LINE MINIMUM 70
CONSTRUCT PERPENDICULAR FROM 63
CONSTRUCT PERPENDICULAR TO 62

CONSTRUCT TANGENT BETWEEN 64
CONSTRUCT TANGENT CIRCLE1 81
CONSTRUCT TANGENT CIRCLE3 80
CONSTRUCT TANGENT FROM 71
CONSTRUCT TANGENT PERPENDICULAR 69
CONSTRUCT TANGENT TO 72
Cutter tools 42

D

Design options 157
Dimensioning 29, 31
Dimensioning Control 181
Dimensioning tools 35
DXF 173
Dynamic Panning 26

E

Edit 17
Element menu 18
Element Symbology Control .... 164 - 165, 167

F

File menu 16
FREEZE 41, 181

H

Hardware 182
Help menu 24

I

IGDS 183
    Differences from MicroStation .... 184 - 185, 187

Similarities to MicroStation 183
Illustrations
   3D tools 46
   MEASURE AREA POINTS 144
   Active color 164
   Active Linestyle 166
   Active Weight 166
   ARRAY POLAR 121
   ARRAY RECTANGULAR 120
   Association lock 37
   Associative 7
   Cancel button 14
   Cell management 160
   CHAMFER 135
   Change Element 167
   Command window 15
   Compatibility 41
   Constant Shading 54
   CONSTRUCT BISECTOR 60
   CONSTRUCT BISECTOR LINE 61
   CONSTRUCT LINE AA1 65
   CONSTRUCT LINE AA2 67
   CONSTRUCT LINE AA3 66
   CONSTRUCT LINE AA4 68
   CONSTRUCT LINE MINIMUM 70
   CONSTRUCT PERPENDICULAR FROM .63
   CONSTRUCT PERPENDICULAR TO 62
   CONSTRUCT TANGENT BETWEEN 64
   CONSTRUCT TANGENT CIRCLE1 81
   CONSTRUCT TANGENT CIRCLE3 80
   CONSTRUCT TANGENT FROM 71
   CONSTRUCT TANGENT PERPENDICULAR 69
   CONSTRUCT TANGENT TO 72
   COPY 115
   Copy Element sub-palette tools 114
   COPY PARALLEL DISTANCE 117
   COPY PARALLEL KEYIN 118
   Crosshatch Area 163
   DELETE PARTIAL 126
   DELETE VERTEX 125
   Design Options 157
   Dimension element tool 40
   Dimension placement 32

Dimension style 33
Dimension Styles 8
Dimension text format 31
Dimensioning Attributes 30
Dimensioning tools 29, 36
DXF configuration 175
DXF Import 174
Edit 17
Element Attribute dialog box 167
EXIT 16
EXTEND LINE 127
EXTEND LINE 2 .... 130 - 131
EXTEND LINE INTERSECTION 129
EXTEND LINE KEYIN 128
FENCE ARRAY POLAR 152
FENCE COPY 150
FENCE DROP 154
FENCE MIRROR COPY VERTICAL 153
Fence pallet tools 146
FENCE ROTATE COPY 151
File 16
FILLET MODIFY 132
FILLET NOMODIFY 133
FILLET SINGLE 134
Fillets sub-palette tools 131
Hatch Area 163
Insert vertex 38, 124
Level Names window 10
Lighting settings box 50
Lines sub-palette tools 105
Main palette 22
Main Tool Palette 156
Match Pattern Attributes 164
MEASURE ANGLE 143
MEASURE AREA ELEMENT 145
MEASURE AREA POINTS 144
MEASURE DISTANCE ALONG 138
MEASURE DISTANCE MINIMUM 141
MEASURE DISTANCE PERPENDICULAR 140
MEASURE DISTANCE POINTS 139
MEASURE RADIUS 142
MicroStation Manager 5, 172
Mirror sub-palette tools 136

MIRROR VERT/HORIZONTAL/LINE 137
MODIFY 123
Modify element 39
MODIFY FENCE 149
Multi-line style 43
Multiline joint tools 42
Multiline style 9
Phong shading 56
Place Active Cell (Interactive) 161
PLACE ARC CENTER 86
PLACE ARC EDGE 87
PLACE ARC RADIUS 88
Place Cell Absolute 161
PLACE CIRCLE CENTER 75
PLACE CIRCLE DIAMETER 76
PLACE CIRCLE EDGE 77
PLACE CIRCLE ISOMETRIC 82
PLACE CIRCLE RADIUS 78
PLACE ELLIPS CENTER 83
PLACE ELLIPSE EDGE 84
PLACE FENCE BLOCK 147
PLACE FENCE SHAPE 148
PLACE LINE 59
PLACE LINE ANGLE 73
Place Line tool 25
PLACE NOTE ....112 - 113
PLACE SLAB 48
PLACE TEXT ....106 - 107
PLACE TEXT ABOVE 108
PLACE TEXT ALONG 110
PLACE TEXT FITTED 111
PLACE TEXT ON 109
plotter options 168
Plotting 168
Renderman Interface 177
ROTATE (ORIGINAL) 116
Selection set 45
Set Page Size 169
Settings menu 19
Smooth shading 55
SPIN ORIGINAL (or copy) 119
sub-palette 26
Text Editor window 107

The Sidebar menu 155
Tool palettes 4
Translation Table 176
View 13
View Rotation 47
Visible edges 179
Working Units 159
IllustrationsSolar lighting 52

L

Lines sub-palette tools 58

M

MDL 5
Multilines 8, 41, 43

N

Named levels 9

O

OSF/Motif GUI 3

P

Palette menu 22
Patterning 163
Phong Shading 55
PLACE ARC CENTER 86
PLACE ARC EDGE 87
PLACE ARC RADIUS 88
PLACE CIRCLE CENTER 75
PLACE CIRCLE DIAMETER 76
PLACE CIRCLE EDGE 77

PLACE CIRCLE ISOMETRIC 82
PLACE CIRCLE RADIUS 78
PLACE ELLIPSE CENTER 83
PLACE ELLIPSE EDGE .... 84 - 85
PLACE LINE 59
PLACE LINE ANGLE 73
PLACE SLAB 47
Plotting .... 168 - 169

R

Reference files 10
Rendering 11, 49, 51, 53, 55
Renderman 177
REPLACE CELL 10

S

SET COMPATIBILITY 41, 181
Settings menu 19
Shading 53
Shared Cells 10
Smooth shading 54

T

Text Attributes 162
THAW 41
Tool Palettes .... 24 - 25
Tool settings 79

U

User menu 23

V

Video considerations 182
View menu 20
View Rotation 46
Visible Edges 178

W

Windows 13
    Stacking 14
Working Units 158

Addenda
MicroStation 4.x Delta Book

By Frank Conforti and David Talbott

Copyright © 1991 Frank Conforti, All Rights Reserved

Addenda
MicroStation 4.x Delta Book

Introduction

Since *MicroStation 4.x Delta Book* was compiled while the new MicroStation was still in beta testing, some new aspects of MicroStation 4.0 were not adequately understood at the time the book went to press. These addenda expand upon the descriptions of a few of MicroStation's important new capabilities.

The Addenda explain in greater detail the Memory Use portion of the Preferences dialog box. They also show how shared cells can be used, how to use the Tool Settings window, the greater flexibility in setting environment variables, and how to use the DOS window.

Allocating System Memory in the Preferences Box

The Preferences settings box, opened by selecting Preferences from the User menu, allows you to set a number of terminal (as opposed to design file) specific preferences. MicroStation saves these settings in a file and reads them each time it starts up.

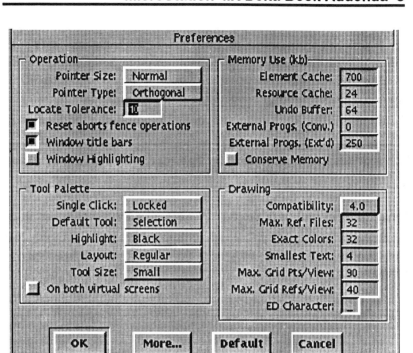

Preferences settings box opened by selecting Preferences from the User menu.

If you are familiar with MicroStation, you will recognize most of the settings in the Preferences box. However, there are a few settings that are entirely new to MicroStation. Most of these are in the Memory Use area of the Preferences box.

NOTE: No changes in the memory usage settings take effect until you exit, clear the handlers from memory and restart MicroStation.

Element Cache

This setting tells how much system memory should be set aside to cache design file data. If your design files are larger than this value, portions of them will be written to temporary files on disk. You can improve system performance by setting this value 10% larger than your typical design file.

Resource Cache

This sets a cache size for caching data read from MicroStation's resource files. The resource cache can speed up the display of palettes and dialog boxes.

This setting is most useful in systems that have no disk caching software. If you have a disk cache already, you can set this value to zero without significantly affecting system performance.

Undo Buffer

This sets the size of the buffer for storing consecutive operations that can be undone with the UNDO command. The larger this buffer, the more operations you can undo.

External Progs, (Conv.)

In MicroStation PC, this sets the amount of conventional memory MicroStation makes available to applications run by shelling out to DOS. This memory is available to applications run with the DOS or ! commands, or applications run through the DOS window with the % command. A setting of -1 allots the maximum amount of available memory.

External Progs, (Ext'd)

This sets the amount of extended memory available to DOS applications run with the DOS, ! or % commands. -1 provides the maximum amount of available memory.

Conserve Memory

When this check button is on, MicroStation conserves RAM by reducing the amount of memory used by the element scanner. Although this will free up some RAM, the added RAM comes at a cost in performance. Unless you have severe RAM restrictions (like you are running MicroStation with only 2 MBytes), you should not set this preference.

Setting Multiple Directories in Environment Variables

A powerful new feature of MicroStation's environment variables is the ability to set more than one directory in a search path. Each directory specification is separated by a semicolon.

For instance, the default setting for the MS_DEF environment variable is MS_DEF=c:\dgn\;\ustation\examples\dgn\. This tells MicroStation to first search c:\dgn\ for a design file whenever you request one. If the requested file is not in c:\dgn, MicroStation will search \ustation\examples\dgn\.

Pop Down Fields and the Tool Settings Window

Many element placement tools can use keyboard input for completion. If you tear off a sub-palette and then select a tool that requires input, pop down fields will appear below the sub-palette.

A more convenient way to key in this information may be through the Tool Settings window. This is a window opened by selecting Tool Settings from the Settings menu, as shown below.

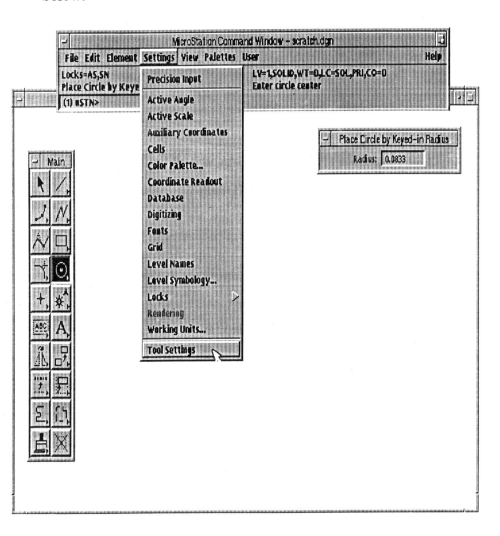

Opening the Tool Settings window.

Any pop down fields pertaining to a command will appear in this box even if you have not torn off the sub-palette. This can be useful if you do not want floating sub-palettes cluttering your screen.

Using the DOS Window

MicroStation PC provides two ways to enter operating system commands without quitting MicroStation.

The DOS or ! commands shell out to DOS and provide you with the operating system prompt. From here you can execute any DOS command or application (memory permitting).

An alternative method of executing DOS applications is through the DOS window. This is a graphic window that opens inside the MicroStation graphic environment. Keying in % *command* opens the DOS window and executes the command.

The advantage of the DOS window is that when an application is done, closing the window only causes a portion of MicroStation's display to redraw. This is faster than returning from the DOS shell and redrawing the entire display.

The DOS window is most useful for quickly calling up a frequently used external applications or operating system commands.

NOTE #1: Only applications that output text through a video BIOS function work in the DOS window. Programs that bypass the video BIOS will not work properly.

NOTE #2: Be sure you have enough memory to run the external application before opening the DOS window with the % command.

Advantages of Shared Cells

The shared cell is a new kind of element introduced with MicroStation 4.0. By keying SET SHARECELL ON (or selecting the Use Shared Cells check button in the Cells settings box), a cell is only defined once in a design file at the time of its initial placement. Each additional occurrence consists only of an element header that defines scale, symbology and properties of the cell.

Shared cells decrease the disk space needed to store a design file. Disk space is conserved because each occurrence of a shared cell does not rewrite all the cell's elements to the design file.

The placement of shared cells happens faster too. This is because there is no cell library scan when additional occurrences of the cell are placed.

For more information on shared cells, see the discussion of shared cells in the *MicroStation 4.X Delta Book*.

More OnWord Press Titles

MicroStation Books

☐ **INSIDE MicroStation**
　　　Book $29.95 Optional Disk $14.95

☐ **INSIDE MicroStation Companion Workbook**
　　　Book $34.95 Includes Disk/Redline Drawings/Projects

☐ **INSIDE MicroStation Companion Workbook Instructor's Guide**
　　　Book $9.95 Includes Disk/Redline Drawings/Projects/Lesson Plans

☐ **MicroStation Reference Guide**
　　　Book $18.95 Optional Disk $14.95

☐ **The MicroStation Productivity Book**
　　　Book $39.95 Optional Disk $49.95

☐ **Programming With MDL**
　　　Book $49.95 Optional Disk $49.95

☐ **Programming With User Commands**
　　　Book $65.00 Optional Disk $40.00

☐ **101 MDL Commands**
　　　　Book $49.95 Optional Executable Disk $101.00

　　　　Optional Source Disks (6) $259.95

☐ **101 User Commands**
　　　　Book $49.95 Optional Disk $101.00

☐ **Bill Steinbock's Pocket MDL Programmers Guide**
　　　　Book $24.95

☐ **MicroStation for AutoCAD Users**
　　　　Book $29.95 Optional Disk $14.95

☐ **MicroStation 4.X Delta Book**
　　　　Book $19.95

☐ **The MicroStation 3D Book**
　　　　Book $39.95 Optional Disk $39.95

☐ **Managing and Networking MicroStation**
　　　　Book $29.95 Optional Disk $29.95

☐ **The MicroStation Database Book**
　　　　Book $29.95 Optional Disk $29.95

☐ **The MicroStation Rendering Book**
　　　　Book $34.95 Includes Disk

☐ **The CLIX Workstation User's Guide**
　　　　Book 39.95

SUNSoft Solaris Series

(Available Fall, 1992)

- ☐ **The SUNSoft Solaris 2.0 User's Guide**
  Book $34.95

- ☐ **The SUNSoft Solaris 2.0 Administrator's Guide**
  Book $34.95

- ☐ **The SUNSoft Solaris 2.0 Quick Reference**
  Book $18.95

- ☐ **Five Steps to SUNSoft Solaris**
  Book $24.95

- ☐ **The One Minute SUNSoft Solaris Manager**
  Book $14.95

The Hewlett Packard HP-UX/HP-VIEW Series

(Available Fall 1992)

- ☐ **The HP-UX/HP-VIEW User's Guide**
  Book $34.95

- ☐ **The HP-UX/HP-VIEW Administrator's Guide**
  Book $34.95

☐ **The HP-UX/HP-VIEW Quick Reference**
    Book $18.95

☐ **Five Steps to HP-UX/HP-VIEW**
    Book $24.95

☐ **The One Minute HP-UX/HP-VIEW Manager**
    Book $14.95

CAD Management

☐ **The One Minute CAD Manager**
    Book $14.95

☐ **The CAD Rating Guide**
    Book $49.00

Geographic Information Systems

☐ **The GIS Book**
    Book $29.95

DTP/CAD Clip Art

☐ **1001 DTP/CAD Symbols Clip Art Library: Architectural**
Book $29.95

    MicroStation .dgn Disk $175.00   Book/Disk $195.00

    AutoCAD .dwg Disk $175.00   Book/Disk $195.00

    CAD/DTP .dxf Disk $195.00   Book/Disk $225.00

    IGES Disk $195.00   Book/Disk $225.00

    TIF Disk $195.00   Book/Disk $225.00

    EPS Disk $195.00   Book/Disk $225.00

    HPGL Disk $195.00   Book/Disk $225.00

    CD ROM With All Formats $275.00   Book/CD $295.00

Networking/Lantastic

☐ **Fantastic Lantastic**
Book $29.95 Includes Disk

☐ **The Lantastic Quick Reference**
Book $14.95

☐ **The One Minute Network Manager**
Book $14.95

OnWord Press Distribution

End User/Corporate
OnWord Press books are available to end users and corporate accounts from your local bookseller or computer/software dealer. Or call in North America: 1-800-223-6397, 505/473-5454, Europe: +44 (0)865 311 361. Fax: U.S. 505/571-4424.

Domestic Trade
OnWord Press books are distributed to the U.S. domestic trade by Publisher's Group West, Emeryville, California.

Domestic Education
OnWord Press books are distributed to the U.S. domestic education markets by Delmar Publishers, Albany, New York.

Wholesale and Site Licensing
Wholesale orders and site licensing call: in North America 800-4 ONWORD or 505/471-2600, Europe: +44 (0)865 311 361. Fax: U.S. 505/538-7171.

OnWord Press, 1580 Center Drive, Santa Fe, NM 87505 USA